The Glow-Worm and Other Beetles

J. Henri Fabre

Translated by Alexander Teixeira de Mattos

Esprios.com

BOOKS BY J. HENRI FABRE

THE LIFE OF THE SPIDER
THE LIFE OF THE FLY
THE MASON-BEES
BRAMBLE-BEES AND OTHERS
THE HUNTING WASPS
THE LIFE OF THE CATERPILLAR
THE LIFE OF THE GRASSHOPPER
THE SACRED BEETLE AND OTHERS
THE MASON-WASPS
THE GLOW-WORM AND OTHER BEETLES

THE GLOW-WORM

AND OTHER BEETLES

BY

J. HENRI FABRE

TRANSLATED BY
ALEXANDER TEIXEIRA DE MATTOS
FELLOW OF THE ZOOLOGICAL SOCIETY OF LONDON

1919

CONTENTS

I	THE GLOW-WORM
II	THE SITARES
III	THE PRIMARY LARVA OF THE SITARES
IV	THE PRIMARY LARVA OF THE OIL-BEETLES
V	HYPERMETAMORPHOSIS
VI	CEROCOMÆ, MYLABRES AND ZONITES
VII	THE CAPRICORN
VIII	THE PROBLEM OF THE SIREX
IX	THE DUNG-BEETLES OF THE PAMPAS
X	INSECT COLOURING
XI	THE BURYING-BEETLES: THE BURIAL
XII	THE BURYING-BEETLES: EXPERIMENTS
XIII	THE GIANT SCARITES
XIV	THE SIMULATION OF DEATH
XV	SUICIDE OR HYPNOSIS?
XVI	THE CRIOCERES
XVII	THE CRIOCERES (*continued*)
XVIII	THE CLYTHRÆ
XIX	THE CLYTHRÆ: THE EGG

TRANSLATOR'S NOTE

This is the second volume on Beetles in the complete English edition of Henri Fabre's entomological works. The first is entitled *The Sacred Beetle and Others*; the second and the third will be known as *The Life of the Weevil* and *More Beetles* respectively.

The Glow-worm, which gives its name to the present book, did not form part of the *Souvenirs entomologiques* as originally published. It is one of two essays written specially, at my request, for translation into English, towards the close of Henri Fabre's life; in fact, this and *The Ant-lion*, a short essay for children, were the last works that came from the veteran author's pen. *The Glow-worm* appeared first in the *Century Magazine*. Of the remaining chapters, several have appeared in various periodicals, notably the *English Review* and in *Land and Water*, the editor and proprietors of which admirable weekly have shown the most enlightened interest in Fabre's work.

A part of the chapter entitled *The Dung-beetles of the Pampas* figures in Messrs. Adam & Charles Black's volume, *The Life and Love of the Insect* (New York: the Macmillan Co.), translated by myself; and the chapters on the Capricorn and Burying-beetles will be found in Mr. T. Fisher Unwin's volume, *The Wonders of Instinct* (New York: the Century Co.), translated by myself and Mr. Bernard Miall, which also contains *The Glow-worm*. These chapters are included in the present edition by consent of and arrangement with the publishers named.

Lastly, Mr. Bernard Miall has earned my gratitude by the valuable assistance which he has given me in preparing the translation of the greater part of this volume.

ALEXANDER TEIXEIRA DE MATTOS.

CHELSEA, *5 September*, 1919.

The Glow-Worm and Other Beetles

CHAPTER I
THE GLOW-WORM

Few insects in our climes vie in popular fame with the Glow-worm, that curious little animal which, to celebrate the little joys of life, kindles a beacon at its tail-end. Who does not know it, at least by name? Who has not seen it roam amid the grass, like a spark fallen from the moon at its full? The Greeks of old called it [Greek: lampouris], meaning, the bright-tailed. Science employs the same term: it calls the lantern-bearer, *Lampyris noctiluca*, LIN. In this case, the common name is inferior to the scientific phrase, which, when translated, becomes both expressive and accurate.

In fact, we might easily cavil at the word "worm." The Lampyris is not a worm at all, not even in general appearance. He has six short legs, which he well knows how to use; he is a gad-about, a trot-about. In the adult state, the male is correctly garbed in wing-cases, like the true Beetle that he is. The female is an ill-favoured thing who knows naught of the delights of flying: all her life long, she retains the larval shape, which, for the rest, is similar to that of the male, who himself is imperfect so long as he has not achieved the maturity that comes with pairing-time. Even in this initial stage, the word "worm" is out of place. We French have the expression "Naked as a worm," to point to the lack of any defensive covering. Now the Lampyris is clothed, that is to say, he wears an epidermis of some consistency; moreover, he is rather richly coloured: his body is dark brown all over, set off with pale pink on the thorax, especially on the lower surface. Finally, each segment is decked at the hinder edge with two spots of a fairly bright red. A costume like this was never worn by a worm.

Let us leave this ill-chosen denomination and ask ourselves what the Lampyris feeds upon. That master of the art of gastronomy, Brillat-Savarin,[1] said:

"Show me what you eat and I will tell you what you are."

The Glow-Worm and Other Beetles

[1] Anthelme Brillat-Savarin (1755-1826), author of *La Psychologie du goût.—Translator's Note*.

A similar question should be addressed, by way of a preliminary, to every insect whose habits we propose to study, for, from the least to the greatest in the zoological progression, the stomach sways the world; the data supplied by food are the chief of all the documents of life. Well, in spite of his innocent appearance, the Lampyris is an eater of flesh, a hunter of game; and he follows his calling with rare villainy. His regular prey is the Snail.

This detail has long been known to entomologists. What is not so well-known, what is not known at all yet, to judge by what I have read, is the curious method of attack, of which I have seen no other instance anywhere.

Before he begins to feast, the Glow-worm administers an anæsthetic: he chloroforms his victim, rivalling in the process the wonders of our modern surgery, which renders the patient insensible before operating on him. The usual game is a small Snail hardly the size of a cherry, such as, for instance, *Helix variabilis*, DRAP., who, in the hot weather, collects in clusters on the stiff stubble and on other long, dry stalks, by the roadside, and there remains motionless, in profound meditation, throughout the scorching summer days. It is in some such resting-place as this that I have often been privileged to light upon the Lampyris banqueting on the prey which he had just paralyzed on its shaky support by his surgical artifices.

But he is familiar with other preserves. He frequents the edges of the irrigating-ditches, with their cool soil, their varied vegetation, a favourite haunt of the mollusc. Here, he treats the game on the ground; and, under these conditions, it is easy for me to rear him at home and to follow the operator's performance down to the smallest detail.

I will try to make the reader a witness of the strange sight. I place a little grass in a wide glass jar. In this I install a few Glow-worms and a provision of Snails of a suitable size, neither too large nor too small, chiefly *Helix variabilis*. We must be patient and wait. Above all, we must keep an assiduous watch, for the desired events come unexpectedly and do not last long.

Here we are at last. The Glow-worm for a moment investigates the prey, which, according to its habit, is wholly withdrawn in the shell, except the edge of the mantle, which projects slightly. Then the hunter's weapon is drawn, a very simple weapon, but one that cannot be plainly perceived without the aid of a lens. It consists of two mandibles bent back powerfully into a hook, very sharp and as thin as a hair. The microscope reveals the presence of a slender groove running throughout the length. And that is all.

The insect repeatedly taps the Snail's mantle with its instrument. It all happens with such gentleness as to suggest kisses rather than bites. As children, teasing one another, we used to talk of "tweaksies" to express a slight squeeze of the finger-tips, something more like a tickling than a serious pinch. Let us use that word. In conversing with animals, language loses nothing by remaining juvenile. It is the right way for the simple to understand one another.

The Lampyris doles out his tweaks. He distributes them methodically, without hurrying, and takes a brief rest after each of them, as though he wished to ascertain the effect produced. Their number is not great: half-a-dozen, at most, to subdue the prey and deprive it of all power of movement. That other pinches are administered later, at the time of eating, seems very likely, but I cannot say anything for certain, because the sequel escapes me. The first few, however—there are never many—are enough to impart inertia and loss of all feeling to the mollusc, thanks to the prompt, I might almost say, lightning methods of the Lampyris, who, beyond a doubt, instils some poison or other by means of his grooved hooks.

Here is the proof of the sudden efficacity of those twitches, so mild in appearance: I take the Snail from the Lampyris, who has operated on the edge of the mantle some four or five times. I prick him with a fine needle in the fore-part, which the animal, shrunk into its shell, still leaves exposed. There is no quiver of the wounded tissues, no reaction against the brutality of the needle. A corpse itself could not give fewer signs of life.

Here is something even more conclusive: chance occasionally gives me Snails attacked by the Lampyris while they are creeping along, the foot slowly crawling, the tentacles swollen to their full extent. A few disordered movements betray a brief excitement on the part of the mollusc and then everything ceases: the foot no longer slugs; the front-part loses its graceful swan-neck curve; the tentacles become limp and give way under their weight, dangling feebly like a broken stick. This conditions persists.

Is the Snail really dead? Not at all, for I am free to resuscitate the seeming corpse. After two or three days of that singular condition which is no longer life and yet not death, I isolate the patient and, although this is not really necessary to success, I give him a douche which will represent the shower so dear to the able-bodied mollusc. In about a couple of days, my prisoner, but lately injured by the Glow-worm's treachery, is restored to his normal state. He revives, in a manner; he recovers movement and sensibility. He is affected by the stimulus of a needle; he shifts his place, crawls, puts out his tentacles, as though nothing unusual had occurred. The general torpor, a sort of deep drunkenness, has vanished outright. The dead returns to life. What name shall we give to that form of existence which, for a time, abolishes the power of movement and the sense of pain? I can see but one that is approximately suitable: anæsthesia. The exploits of a host of Wasps whose flesh-eating grubs are provided with meat that is motionless though not dead[2] have taught us the skilful art of the paralyzing insect, which numbs the locomotory nerve-centres with its venom. We have now a humble little animal that first produces complete anæsthesia in its patient. Human science did not in reality invent this art, which is one of the wonders of our latter-day surgery. Much earlier, far back in the centuries, the Lampyris and, apparently, others knew it as well. The animal's knowledge had a long start of ours; the method alone has changed. Our operators proceed by making us inhale the fumes of ether or chloroform; the insect proceeds by injecting a special virus that comes from the mandibular fangs in infinitesimal doses. Might we not one day be able to benefit by this hint? What glorious discoveries the future would have in store for us, if we understood the beastie's secrets better!

² Cf. *The Hunting Wasps*, by J. Henri Fabre, translated by Alexander Teixeira de Mattos: *passim.—Translator's Note*.

What does the Lampyris want with anæsthetical talent against a harmless and moreover eminently peaceful adversary, who would never begin the quarrel of his own accord? I think I see. We find in Algeria a Beetle known as *Drilus maroccanus*, who, though non-luminous, approaches our Glow-worm in his organization and especially in his habits. He too feeds on land molluscs. His prey is a Cyclostome with a graceful spiral shell, tight-closed with a stony lid which is attached to the animal by a powerful muscle. The lid is a movable door which is quickly shut by the inmate's mere withdrawal into his house and as easily opened when the hermit goes forth. With this system of closing, the abode becomes inviolable; and the Drilus knows it.

Fixed to the surface of the shell by an adhesive apparatus whereof the Lampyris will presently show us the equivalent, he remains on the look-out, waiting, if necessary, for whole days at a time. At last, the need of air and food oblige the besieged noncombatant to show himself; at least, the door is set slightly ajar. That is enough. The Drilus is on the spot and strikes his blow. The door can no longer be closed and the assailant is henceforth master of the fortress. Our first impression is that the muscle moving the lid has been cut with a quick-acting pair of shears. This idea must be dismissed. The Drilus is not well enough equipped with jaws to gnaw through a fleshy mass so promptly. The operation has to succeed at once, at the first touch: if not, the animal attacked would retreat, still in full vigour, and the siege must be recommenced, as arduous as ever, exposing the insect to fasts indefinitely prolonged. Although I have never come across the Drilus, who is a stranger to my district, I conjecture a method of attack very similar to that of the Glow-worm. Like our own Snail-eater, the Algerian insect does not cut its victim into small pieces: it renders it inert, chloroforms it by means of a few tweaks which are easily distributed, if the lid but half-opens for a second. That will do. The besieger thereupon enters and, in perfect quiet, consumes a prey incapable of the least muscular effort. That is how I see things by the unaided light of logic.

Let us now return to the Glow-worm. When the Snail is on the ground, creeping, or even shrunk into his shell, the attack never presents any difficulty. The shell possesses no lid and leaves the hermit's fore-part to a great extent exposed. Here, on the edges of the mantle contracted by the fear of danger, the mollusc is vulnerable and incapable of defence. But it also frequently happens that the Snail occupies a raised position, clinging to the tip of a grass-stalk or perhaps to the smooth surface of a stone. This support serves him as a temporary lid; it wards off the aggression of any churl who might try to molest the inhabitant of the cabin, always on the express condition that no slit show itself anywhere on the protecting circumference. If, on the other hand, in the frequent case when the shell does not fit its support quite closely, some point, however tiny, be left uncovered, this is enough for the subtle tools of the Lampyris, who just nibbles at the mollusc and at once plunges him into that profound immobility which favours the tranquil proceedings of the consumer.

These proceedings are marked by extreme prudence. The assailant has to handle his victim gingerly, without provoking contractions which would make the Snail let go his support and, at the very least, precipitate him from the tall stalk whereon he is blissfully slumbering. Now any game falling to the ground would seem to be so much sheer loss, for the Glow-worm has no great zeal for hunting-expeditions: he profits by the discoveries which good luck sends him, without undertaking assiduous searches. It is essential, therefore, that the equilibrium of a prize perched on the top of a stalk and only just held in position by a touch of glue should be disturbed as little as possible during the onslaught; it is necessary that the assailant should go to work with infinite circumspection and without producing pain, lest any muscular reaction should provoke a fall and endanger the prize. As we see, sudden and profound anæsthesia is an excellent means of enabling the Lampyris to attain his object, which is to consume his prey in perfect quiet.

What is his manner of consuming it? Does he really eat, that is to say, does he divide his food piecemeal, does he carve it into minute particles, which are afterwards ground by a chewing-apparatus? I think not. I never see a trace of solid nourishment on my captives' mouths. The Glow-worm does not eat in the strict sense of the word: he drinks his fill; he feeds on a thin gruel into which he transforms his prey by a method recalling that of the maggot. Like the flesh-eating grub of the Fly, he too is able to digest before consuming; he liquefies his prey before feeding on it.

This is how things happen: a Snail has been rendered insensible by the Glow-worm. The operator is nearly always alone, even when the prize is a large one, like the Common Snail, *Helix aspersa*. Soon a number of guests hasten up—two, three or more—and, without any quarrel with real proprietor, all alike fall to. Let us leave them to themselves for a couple of days and then turn the shell, with the opening downwards. The contents flow out as easily as would soup from an overturned saucepan. When the sated diners retire from this gruel, only insignificant leavings remain.

The matter is obvious: by repeated tiny bites, similar to the tweaks which we saw distributed at the outset, the flesh of the mollusc is converted into a gruel on which the various banqueters nourish themselves without distinction, each working at the broth by means of some special pepsine and each taking his own mouthfuls of it. In consequence of this method, which first converts the food into a liquid, the Glow-worm's mouth must be very feebly armed apart from the two fangs which sting the patient and inject the anæsthetic poison and, at the same time, no doubt, the serum capable of turning the solid flesh into fluid. These two tiny implements, which can just be examined through the lens, must, it seems, have some other object. They are hollow and in this resemble those of the Ant-lion, which sucks and drains its capture without having to divide it; but there is this great difference, that the Ant-lion leaves copious remnants, which are afterwards flung outside the funnel-shaped trap dug in the sand, whereas the Glow-worm, that expert liquefier, leaves nothing, or next to nothing. With similar tools, the one simply sucks the blood of its prey and the other turns every morsel of his to account, thanks to a preliminary liquefaction.

And this is done with exquisite precision, though the equilibrium is sometimes anything but steady. My rearing-glasses supply me with magnificent examples. Crawling up the sides, the Snails imprisoned in my apparatus sometimes reach the top, which is closed with a glass pane, and fix themselves to it by means of a speck of glair. This is a mere temporary halt, in which the mollusc is miserly with its adhesive product, and the merest shake is enough to loosen the shell and send it to the bottom of the jar.

Now it is not unusual for the Glow-worm to hoist himself to the top, with the help of a certain climbing-organ that makes up for his weak legs. He selects his quarry, makes a minute inspection of it to find an entrance-slit, nibbles it a little, renders it insensible and, without delay, proceeds to prepare the gruel which he will consume for days on end.

When he leaves the table, the shell is found to be absolutely empty; and yet this shell, which was fixed to the glass by a very faint stickiness, has not come loose, has not even shifted its position in the smallest degree: without any protest from the hermit gradually converted into broth, it has been drained on the very spot at which the first attack was delivered. These small details tell us how promptly the anæsthetic bite takes effect; they teach us how dexterously the Glow-worm treats his Snail without causing him to fall from a very slippery vertical support and without even shaking him on his slight line of adhesion.

Under these conditions of equilibrium, the operator's short, clumsy legs are obviously not enough; a special accessory apparatus is needed to defy the danger of slipping and to seize the unseizable. And this apparatus the Lampyris possesses. At the hinder end of the animal we see a white spot which the lens separates into some dozen short, fleshy appendages, sometimes gathered into a cluster, sometimes spread into a rosette. There is your organ of adhesion and locomotion. If he would fix himself somewhere, even on a very smooth surface, such as a grass-stalk, the Glow-worm opens his rosette and spreads it wide on the support, to which it adheres by its own stickiness. The same organ, rising and falling, opening and closing, does much to assist the act of progression. In short, the Glow-worm is a new sort of self-propelled cripple, who decks his hind-quarters with a dainty white rose, a kind of hand with twelve fingers, not jointed, but moving in every direction: tubular fingers which do not seize, but stick.

The same organ serves another purpose: that of a toilet-sponge and brush. At a moment of rest, after a meal, the Glow-worm passes and repasses the said brush over his head, back, sides and hinder-parts, a performance made possible by the flexibility of his spine. This is done point by point, from one end of the body to the other, with a scrupulous persistency that proves the great interest which he takes in the operation. What is his object in thus sponging himself, in dusting and polishing himself so carefully? It is a question, apparently, of removing a few atoms of dust or else some traces of viscidity that remain from the evil contact with the snail. A wash and brush-up is not superfluous when one leaves the tub in which the mollusc has been treated.

If the Glow-worm possessed no other talent than that of chloroforming his prey by means of a few tweaks resembling kisses, he would be unknown to the vulgar herd; but he also knows how to light himself like a beacon; he shines, which is an excellent manner of achieving fame. Let us consider more particularly the female, who, while retaining her larval shape, becomes marriageable and glows at her best during the hottest part of summer. The lighting-apparatus occupies the last three segments of the abdomen. On each of the first two, it takes the form, on the ventral surface, of a wide belt covering almost the whole of the arch; on the third, the luminous part is much less and consists simply of two small crescent-shaped markings, or rather two spots which shine through to the back and are visible both above and below the animal. Belts and spots emit a glorious white light, delicately tinged with blue. The general lighting of the

Glow-worm thus comprises two groups: first, the wide belts of the two segments preceding the last; secondly, the two spots of the final segments. The two belts, the exclusive attribute of the marriageable female, are the part richest in light: to glorify her wedding, the future mother dons her brightest gauds; she lights her two resplendent scarves. But, before that, from the time of the hatching, she had only the modest rush-light of the stern. This efflorescence of light is the equivalent of the final metamorphosis, which is usually represented by the gift of wings and flight. Its brilliance heralds the pairing-time. Wings and flight there will be none: the female retains her humble larval form, but she kindles her blazing beacon.

The male, on his side, is fully transformed, changes his shape, acquires wings and wing-cases; nevertheless, like the female, he possesses, from the time when he is hatched, the pale lamp of the end segment. This luminous aspect of the stern is characteristic of the entire Glow-worm tribe, independently of sex and season. It appears upon the budding grub and continues throughout life unchanged. And we must not forget to add that it is visible on the dorsal as well as on the ventral surface, whereas the two large belts peculiar to the female shine only under the abdomen.

My hand is not so steady nor my sight so good as once they were, but, as far as they allow me, I consult anatomy for the structure of the luminous organs. I take a scrap of the epidermis and manage to separate pretty neatly half of one of the shining belts. I place my preparation under the microscope. On the skin, a sort of white-wash lies spread, formed of a very fine, granular substance. This is certainly the light-producing matter. To examine this white layer more closely is beyond the power of my weary eyes. Just beside it is a curious air-tube, whose short and remarkably wide stem branches suddenly into a sort of bushy tuft of very delicate ramifications. These creep over the luminous sheet, or even dip into it. That is all.

The luminescence, therefore, is controlled by the respiratory organs and the work produced is an oxidization. The white sheet supplies the oxidizable matter and the thick air-tube spreading into a tufty bush distributes the flow of air over it. There remains the question of the substance whereof this sheet is formed. The first suggestion was phosphorus, in the chemist's sense of the word. The Glow-worm has been calcined and treated with the violent reagents that bring the simple substances to light; but no one, so far as I know, has obtained a satisfactory answer along these lines. Phosphorus seems to play no part here, in spite of the name of phosphorescence which is sometimes bestowed upon the Glow-worm's gleam. The answer lies elsewhere, no one knows where.

We are better informed as regards another question. Has the Glow-worm a free control of the light which he emits? Can he turn it on or down or put it out as he pleases? Has he an opaque screen which is drawn over the flame at will, or is that flame always left exposed? There is no need for any such mechanism: the insect has something better for its revolving light.

The thick tube supplying the light-producing sheet increases the flow of air and the light is intensified; the same air-tube, swayed by the animal's will, slackens or even suspends the passage of air and the light grows fainter or even goes out. It is, in short, the mechanism of a lamp which is regulated by the access of air to the wick.

Excitement can set the attendant air-duct in motion. We must here distinguish between two cases: that of the gorgeous scarves, the exclusive ornament of the female ripe for matrimony, and that of the modest fairy-lamp on the last segment, which both sexes kindle at any age. In the second case, the extinction caused by a flurry is sudden and complete, or nearly so. In my nocturnal hunts for young Glow-worms, measuring about 5 millimetres long,[3] I can plainly see the glimmer on the blades of grass; but, should the least false step disturb a neighbouring twig, the light goes out at once and the coveted insect becomes invisible. Upon the full-grown females, lit up with their nuptial scarves, even a violent start has but a slight effect and often none at all.

[3] .195 inch.—*Translator's Note.*

The Glow-Worm and Other Beetles

I fire a gun beside a wire-gauze cage in which I am rearing my menagerie of females in the open air. The explosion produces no result. The illumination continues, as bright and placid as before. I take a spray and rain down a slight shower of cold water upon the flock. Not one of my animals puts out its light; at the very most, there is a brief pause in the radiance; and then only in some cases. I send a puff of smoke from my pipe into the cage. This time, the pause is more marked. There are even some extinctions, but these do not last long. Calm soon returns and the light is renewed as brightly as ever. I take some of the captives in my fingers, turn and return them, tease them a little. The illumination continues and is not much diminished, if I do not press too hard with my thumb. At this period, with the pairing close at hand, the insect is in all the fervour of its passionate splendour; and nothing short of very serious reasons would make it put out its signals altogether.

All things considered, there is not a doubt but that the Glow-worm himself manages his lighting-apparatus, extinguishing and rekindling it at will; but there is one point at which the voluntary agency of the insect is without effect. I detach a strip of the epidermis showing one of the luminescent sheets and place it in a glass tube, which I close with a plug of damp wadding, to avoid too rapid an evaporation. Well, this scrap of carcass shines away merrily, although not quite as brilliantly as on the living body.

Life's aid is now superfluous. The oxidizable substance, the luminescent sheet, is in direct communication with the surrounding atmosphere; the flow of oxygen through an air-tube is not necessary; and the luminous emission continues to take place, in the same way as when it is produced by the contact of the air with the real phosphorus of the chemists. Let us add that, in aerated water, the luminousness continues as brilliant as in the free air, but that it is extinguished in water deprived of its air by boiling. No better proof could be found of what I have already propounded, namely, that the Glow-worm's light is the effect of a slow oxidization.

The light is white, calm and soft to the eyes and suggests a spark dropped by the full moon. Despite its splendour, it is a very feeble illuminant. If we move a Glow-worm along a line of print, in perfect darkness, we can easily make out the letters, one by one, and even words, when these are not too long; but nothing more is visible beyond a narrow zone. A lantern of this kind soon tires the reader's patience.

Suppose a group of Glow-worms placed almost touching one another. Each of them sheds its glimmer, which ought, one would think, to light up its neighbours by reflexion and give us a clear view of each individual specimen. But not at all: the luminous party is a chaos in which our eyes are unable to distinguish any definite form at a medium distance. The collective lights confuse the link-bearers into one vague whole.

Photography gives us a striking proof of this. I have a score of females, all at the height of their splendour, in a wire-gauze cage in the open air. A tuft of thyme forms a grove in the centre of their establishment. When night comes, my captives clamber to this pinnacle and strive to show off their luminous charms to the best advantage at every point of the horizon, thus forming along the twigs marvellous clusters from which I expected magnificent effects on the photographer's plates and paper. My hopes are disappointed. All that I obtain is white, shapeless patches, denser here and less dense there according to the numbers forming the group. There is no picture of the Glow-worms themselves; not a trace either of the tuft of thyme. For want of satisfactory light, the glorious firework is represented by a blurred splash of white on a black ground.

The beacons of the female Glow-worms are evidently nuptial signals, invitations to the pairing; but observe that they are lighted on the lower surface of the abdomen and face the ground, whereas the summoned males, whose flights are sudden and uncertain, travel overhead, in the air, sometimes a great way up. In its normal position, therefore, the glittering lure is concealed from the eyes of those concerned; it is covered by the thick bulk of the bride. The lantern ought really to gleam on the back and not under the belly; otherwise the light is hidden under a bushel.

The Glow-Worm and Other Beetles

The anomaly is corrected in a very ingenious fashion, for every female has her little wiles of coquetry. At nightfall, every evening, my caged captives make for the tuft of thyme with which I have thoughtfully furnished the prison and climb to the top of the upper branches, those most in sight. Here, instead of keeping quiet, as they did at the foot of the bush just now, they indulge in violent exercises, twist the tip of their very flexible abdomen, turn it to one side, turn it to the other, jerk it in every direction. In this way, the search-light cannot fail to gleam, at one moment or another, before the eyes of every male who goes a-wooing in the neighbourhood, whether on the ground or in the air.

It is very like the working of the revolving mirror used in catching Larks. If stationary, the little contrivance would leave the bird indifferent; turning and breaking up its light in rapid flashes, it excites it.

While the female Glow-worm has her tricks for summoning her swains, the male, on his side, is provided with an optical apparatus suited to catch from afar the least reflection of the calling-signal. His corselet expands into a shield and overlaps his head considerably in the form of a peaked cap or eye-shade, the object of which appears to be to limit the field of vision and concentrate the view upon the luminous speck to be discerned. Under this arch are the two eyes, which are relatively enormous, exceedingly convex, shaped like a skull-cap and contiguous to the extent of leaving only a narrow groove for the insertion of the antennæ. This double eye, occupying almost the whole face of the insect and contained in the cavern formed by the spreading peak of the corselet, is a regular Cyclop's eye.

At the moment of the pairing, the illumination becomes much fainter, is almost extinguished; all that remains alight is the humble fairy-lamp of the last segment. This discreet night-light is enough for the wedding, while, all around, the host of nocturnal insects, lingering over their respective affairs, murmur the universal marriage-hymn. The laying follows very soon. The round, white eggs are laid, or rather strewn at random, without the least care on the mother's part, either on the more or less cool earth or on a blade of grass. These brilliant ones know nothing at all of family-affection.

Here is a very singular thing: the Glow-worm's eggs are luminous even when still contained in the mother's womb. If I happen by accident to crush a female big with germs that have reached maturity, a shiny streak runs along my fingers, as though I had broken some vessel filled with a phosphorescent fluid. The lens shows me that I am wrong. The luminosity comes from the cluster of eggs forced out of the ovary. Besides, as laying-time approaches, the phosphorescence of the eggs is already made manifest without this clumsy midwifery. A soft opalescent light shines through the skin of the belly.

The hatching follows soon after the laying. The young of either sex have two little rush-lights on the last segment. At the approach of the severe weather, they go down into the ground, but not very far. In my rearing-jars, which are supplied with fine and very loose earth, they descend to a depth of three or four inches at most. I dig up a few in mid-winter. I always find them carrying their faint stern-light. About the month of April, they come up again to the surface, there to continue and complete their evolution.

From start to finish, the Glow-worm's life is one great orgy of light. The eggs are luminous; the grubs likewise. The full-grown females are magnificent light-houses, the adult males retain the glimmer which the grubs already possessed. We can understand the object of the feminine beacon; but of what use is all the rest of the pyrotechnic display? To my great regret, I cannot tell. It is and will be, for many a day to come, perhaps for all time, the secret of animal physics, which is deeper than the physics of the books.

CHAPTER II
THE SITARES

The high banks of sandy clay in the country round about Carpentras are the favourite haunts of a host of Bees and Wasps, those lovers of a thoroughly sunny aspect and of soils that are easy to excavate. Here, in the month of May, two Anthophoræ[1] are especially abundant, gatherers of honey and, both of them, makers of subterranean cells. One, *A. parietina,* builds at the entrance of her dwelling an advanced fortification, an earthy cylinder, wrought in open work, like that of the Odynerus,[2] and curved like it, but of the width and length of a man's finger. When the community is a populous one, we stand amazed at the rustic ornamentation formed by all these stalactites of clay hanging from the façade. The other, *A. pilipes,* who is very much more frequent, leaves the opening of her corridor bare. The chinks between the stones in old walls and abandoned hovels, the surfaces of excavations in soft sandstone or marl, are found suitable for her labours; but the favourite spots, those to which the greatest number of swarms resort, are vertical stretches, exposed to the south, such as are afforded by the cuttings of deeply sunken roads. Here, over areas many yards in width, the wall is drilled with a multitude of holes, which impart to the earthy mass the look of some enormous sponge. These round holes might be fashioned with an auger, so regular are they. Each is the entrance to a winding corridor, which runs to a depth of four to six inches. The cells are distributed at the far end. If we would witness the labours of the industrious Bee, we must repair to her workshop during the latter half of May. Then, but at a respectful distance, if, as novices, we are afraid of being stung, we may contemplate, in all its bewildering activity, the tumultuous, buzzing swarm, busied with the building and the provisioning of the cells.

[1] Cf. *The Mason-bees,* by J. Henri Fabre, translated by Alexander Teixeira de Mattos: chap. viii.; and *Bramble-bees and Others,* by J. Henri Fabre, translated by Alexander Teixeira de Mattos: *passim.*—*Translator's Note.*

[2] Cf. *The Mason-wasps*, by J. Henri Fabre, translated by Alexander Teixeira de Mattos: chaps. vi. and x.—*Translator's Note*.

It is most often during the months of August and September, those happy months of the summer holidays, that I have visited the banks inhabited by the Anthophora. At this period all is silent near the nests; the work has long been completed; and numbers of Spiders' webs line the crevices or plunge their silken tubes into the Bee's corridors. Let us not, however, hastily abandon the city once so populous, so full of life and bustle and now deserted. A few inches below the surface, thousands of larvæ and nymphs, imprisoned in their cells of clay, are resting until the coming spring. Might not such a succulent prey as these larvæ, paralysed and incapable of defence, tempt certain parasites who are industrious enough to attain them?

Here indeed are some Flies clad in a dismal livery, half-black, half-white, a species of Anthrax (*A. sinuata*),[3] flying indolently from gallery to gallery, doubtless with the object of laying their eggs there; and here are others, more numerous, whose mission is fulfilled and who, having died in harness, are hanging dry and shrivelled in the Spiders' webs. Elsewhere the entire surface of a perpendicular bank is hung with the dried corpses of a Beetle (*Sitaris humeralis*), slung, like the Flies, in the silken meshes of the Spiders. Among these corpses some male Sitares circle, busy, amorous, heedless of death, mating with the first female that passes within reach, while the fertilized females thrust their bulky abdomens into the opening of a gallery and disappear into it backwards. It is impossible to mistake the situation: some grave interest attracts to this spot these two insects, which, within a few days, make their appearance, mate, lay their eggs and die at the very doors of the Anthophora's dwellings.

[3] Cf. *The Life of the Fly*, by J. Henri Fabre, translated by Alexander Teixeira de Mattos: chaps. ii. and iv.—*Translator's Note*.

Let us now give a few blows of the pick to the surface beneath which the singular incidents already in our mind must be occurring, beneath which similar things occurred last year; perhaps we shall find some evidence of the parasitism which we suspected. If we search the dwellings of the Anthophoræ during the early days of August, this is what we see: the cells forming the superficial layer are not like those situated at a greater depth. This difference arises from the fact that the same establishment is exploited simultaneously by the Anthophora and by an Osmia (*O. tricornis*)[4] as is proved by an observation made at the working-period, in May. The Anthophoræ are the actual pioneers, the work of boring the galleries is wholly theirs; and their cells are situated right at the end. The Osmia profits by the galleries which have been abandoned either because of their age, or because of the completion of the cells occupying the most distant part; she builds her cells by dividing these corridors into unequal and inartistic chambers by means of rude earthen partitions. The Osmia's sole achievement in the way of masonry is confined to these partitions. This, by the way, is the ordinary building-method adopted by the various Osmiæ, who content themselves with a chink between two stones, an empty Snail-shell, or the dry and hollow stem of some plant, wherein to build their stacks of cells, at small expense, by means of light partitions of mortar.

[4] Cf. *Bramble-bees and Others: passim.*—Translator's Note.

The cells of the Anthophora, with their faultless geometrical regularity and their perfect finish, are works of art, excavated, at a suitable depth, in the very substance of the loamy bank, without any manufactured part save the thick lid that closes the orifice. Thus protected by the prudent industry of their mother, well out of reach in their distant, solid retreats, the Anthophora's larvæ are devoid of the glandular apparatus designed for secreting silk. They therefore never spin a cocoon, but lie naked in their cells, whose inner surface has the polish of stucco.

In the Osmia's cells, on the other hand, means of defence are required, for these are situated in the surface layer of the bank; they are irregular in form, rough inside and barely protected, by their thin earthen partitions, against external enemies. The Osmia's larvæ, in fact, contrive to enclose themselves in an egg-shaped cocoon, dark brown in colour and very strong, which preserves them both from the rough contact of their shapeless cells and from the mandibles of voracious parasites, Acari,[5] Cleri[6] and Anthreni,[7] those manifold enemies whom we find prowling in the galleries, seeking whom they may devour. It is by means of this equipoise between the mother's talents and the larva's that the Osmia and the Anthophora, in their early youth, escape some part of the dangers which threaten them. It is easy therefore, in the bank excavated by these two Bees, to recognize the property of either species by the situation and form of the cells and also by their contents, which consist, with the Anthophora, of a naked larva and, with the Osmia, of a larva enclosed in a cocoon.

[5] Mites and Ticks.—*Translator's Note.*

[6] A genus of Beetles of which certain species (*Clerus apiarius* and *C. alvearius*) pass their preparatory state in the nests of Bees, where they feed on the grubs.—*Translator's Note.*

[7] Another genus of Beetles. The grub of *A. musæorum*, the Museum Beetle, is very destructive to insect-collections.—*Translator's Note.*

On opening a certain number of these cocoons, we end by discovering some which, in place of the Osmia's larva, contain each a curiously shaped nymph. These nymphs, at the least shock received by their dwelling, indulge in extravagant movements, lashing the walls with their abdomen till the whole house shakes and dances. And, even if we leave the cocoon intact, we are informed of their presence by a dull rustle heard inside the silken dwelling the moment after we move it.

The fore-part of this nymph is fashioned like a sort of boar's-snout armed with six strong spikes, a multiple ploughshare, eminently adapted for burrowing in the soil. A double row of hooks surmounts the dorsal ring of the four front segments of the abdomen. These are so many grappling-irons, with whose assistance the creature is enabled to progress in the narrow gallery dug by the snout. Lastly, a sheaf of sharp points forms the armour of the hinder-part. If we examine attentively the surface of the vertical wall which contains the various nests, it will not be long before we discover nymphs like those which we have been describing, with one extremity held in a gallery of their own diameter, while the fore-part projects freely into the air. But these nymphs are reduced to their cast skins, along the back and head of which runs a long slit through which the perfect insect has escaped. The purpose of the nymph's powerful weapons is thus made manifest: it is the nymph that has to rend the tough cocoon which imprisons it, to excavate the tightly-packed soil in which it is buried, to dig a gallery with its six-pointed snout and thus to bring to the light the perfect insect, which apparently is incapable of performing these strenuous tasks for itself.

And in fact these nymphs, taken in their cocoons, have in a few days' time given me a feeble Fly (*Anthrax sinuata*) who is quite incapable of piercing the cocoon and still more of making her exit through a soil which I cannot easily break up with my pick. Although similar facts abound in insect history, we always notice them with a lively interest. They tell us of an incomprehensible power which suddenly, at a given moment, irresistibly commands an obscure grub to abandon the retreat in which it enjoys security, in order to make its way through a thousand difficulties and to reach the light, which would be fatal to it on any other occasion, but which is necessary to the perfect insect, which could not reach it by its own efforts.

But the layer of Osmia-cells has been removed; and the pick now reaches the Anthophora's cells. Among these cells are some which contain larvæ and which result from the labours of last May; others, though of the same date, are already occupied by the perfect insect. The precocity of metamorphosis varies from one larva to another; however, a few days' difference of age is enough to explain these inequalities of development. Other cells, as numerous as the first, contain a parasitical Hymenopteron, a Melecta (*M. armata*), likewise in the perfect state. Lastly, there are some, indeed many, which contain a singular egg-shaped shell, divided into segments with projecting breathing-pores. This shell is extremely thin and fragile; it is amber-coloured and so transparent that one can distinguish quite plainly, through its sides, an adult Sitaris (*S. humeralis*), who occupies the interior and is struggling as though to set herself at liberty. This explains the presence here, the pairing and the egg-laying of the Sitares whom we but now saw roaming, in the company of the Anthrax-flies, at the entrance to the galleries of the Anthophoræ. The Osmia and the Anthophora, the joint owners of the premises, have each their parasite: the Anthrax attacks the Osmia and the Sitaris the Anthophora.

But what is this curious shell in which the Sitaris is invariably enclosed, a shell unexampled in the Beetle order? Can this be a case of parasitism in the second degree, that is, can the Sitaris be living inside the chrysalis of a first parasite, which itself exists at the cost of the Anthophora's larva or of its provisions? And, even so, how can this parasite, or these parasites, obtain access to a cell which seems to be inviolable, because of the depth at which it lies, and which, moreover, does not reveal, to the most careful examination under the magnifying-glass, any violent inroad on the enemy's part? These are the questions that presented themselves to my mind when for the first time, in 1855, I observed the facts which I have just related. Three years of assiduous observation enabled me to add one of its most astonishing chapters to the story of the formation of insects.

After collecting a fairly large number of these enigmatical shells containing adult Sitares, I had the satisfaction of observing, at leisure, the emergence of the perfect insect from the shell, the act of pairing and the laying of the eggs. The shell is easily broken; a few strokes of the mandibles, distributed at random, a few kicks are enough to deliver the perfect insect from its fragile prison.

In the glass jars in which I kept my Sitares I saw the pairing follow very closely upon the first moments of freedom. I even witnessed a fact which shows emphatically how imperious, in the perfect insect, is the need to perform, without delay, the act intended to ensure the preservation of its race. A female, with her head already cut out of the shell, is anxiously struggling to release herself entirely; a male, who has been free for a couple of hours, climbs on the shell and, tugging here and there, with his mandibles, at the fragile envelope, strives to deliver the female from her shackles. His efforts are soon crowned with success; and, though the female is still three parts swathed in her swaddling-bands, the coupling takes place immediately, lasting about a minute. During the act, the male remains motionless on the top of the shell, or on the top of the female when the latter is entirely free. I do not know whether, in ordinary circumstances, the male occasionally thus helps the female to gain her liberty; to do so he would have to penetrate into a cell containing a female, which, after all, is not beyond his powers, seeing that he has been able to escape from his own. Still, on the actual site of the cells, the coupling is generally performed at the entrance to the galleries of the Anthophoræ; and then neither of the sexes drags about with it the least shred of the shell from which it has emerged.

After mating, the two Sitares proceed to clean their legs and antennæ by drawing them between their mandibles; then each goes his own way. The male cowers in a crevice of the earthen bank, lingers for two or three days and perishes. The female also, after getting rid of her eggs, which she does without delay, dies at the entrance to the corridor in which the eggs are laid. This is the origin of all those corpses swinging in the Spiders' web with which the neighbourhood of the Anthophora's dwellings is upholstered.

Thus the Sitares in the perfect state live long enough only to mate and to lay their eggs. I have never seen one save upon the scene of their loves, which is also that of their death; I have never surprised one browsing on the plants near at hand, so that, though they are provided with a normal digestive apparatus, I have grave reasons to doubt whether they actually take any nourishment whatever. What a life is theirs! A fortnight's feasting in a storehouse of honey; a year of slumber underground; a minute of love in the sunlight; then death!

Once fertilized, restlessly the female at once proceeds to seek a favourable spot wherein to lay her eggs. It was important to note where this exact spot is. Does the female go from cell to cell, confiding an egg to the succulent flanks of each larva, whether this larva belong to the Anthophora or to a parasite of hers, as the mysterious shell whence the Sitaris emerges would incline one to believe? This method of laying the eggs, one at a time in each cell, would appear to be essential, if we are to explain the facts already ascertained. But then why do the cells usurped by the Sitares retain not the slightest trace of the forcible entry which is indispensable? And how is it that, in spite of lengthy investigations during which my perseverance has been kept up by the keenest desire to cast some light upon all these mysteries, how is it, I say, that I have never come across a single specimen of the supposed parasites to which the shell might be attributed, since this shell appears not to be a Beetle's? The reader would hardly suspect how my slight acquaintance with entomology was unsettled by this inextricable maze of contradictory facts. But patience! We may yet obtain some light.

Let us begin by observing precisely at what spot the eggs are laid. A female has just been fertilized before my eyes; she is forthwith placed in a large glass jar, into which I put, at the same time, some clods of earth containing Anthophora-cells. These cells are occupied partly by larvæ and partly by nymphs that are still quite white; some are slightly open and afford a glimpse of their contents. Lastly, in the inner surface of the cork which closes the jar I sink a cylindrical well, a blind alley, of the same diameter as the corridors of the Anthophora. In order that the insect, if it so desire, may enter this artificial corridor, I lay the bottle horizontally.

The female, painfully dragging her big abdomen, perambulates all the nooks and corners of her makeshift dwelling, exploring them with her palpi, which she passes everywhere. After half an hour of groping and careful investigation, she ends by selecting the horizontal gallery dug in the cork. She thrusts her abdomen into this cavity and, with her head hanging outside, begins her laying. Not until thirty-six hours later was the operation completed; and during this incredible lapse of time the patient creature remained absolutely motionless.

The eggs are white, oval and very small. They measure barely two-thirds of a millimetre[8] in length. They stick together slightly and are piled in a shapeless heap which might be likened to a good-sized pinch of the unripe seeds of some orchid. As for their number, I will admit that it tried my patience to no purpose. I do not, however, believe that I am exaggerating when I estimate it as at least two thousand. Here are the data on which I base this figure: the laying, as

I have said, lasts thirty-six hours; and my frequent visits to the female working in the cavity in the cork convinced me that there was no perceptible interruption in the successive emission of the eggs. Now less than a minute elapses between the arrival of one egg and that of the next; and the number of these eggs cannot therefore be lower than the number of minutes contained in thirty-six hours, or 2160. But the exact number is of no importance: we need only note that it is very large, which implies, for the young larvæ issuing from the eggs, very numerous chances of destruction, since so lavish a supply of germs is necessary to maintain the species in the requisite proportions.

[8] .026 inch.—*Translator's Note.*

Enlightened by these observations and informed of the shape, the number and the arrangement of the eggs, I searched the galleries of the Anthophoræ for those which the Sitares had laid there and invariably found them gathered in a heap inside the galleries, at a distance of an inch or two from the orifice, which is always open to the outer world. Thus, contrary to what one was to some extent entitled to suppose, the eggs are not laid in the cells of the pioneer Bee; they are simply dumped in a heap inside the entrance to her dwelling. Nay more, the mother does not make any protective structure for them; she takes no pains to shield them from the rigours of winter; she does not even attempt, by stopping for a short distance, as best she can, the entrance-lobby in which she has laid them, to protect them from the thousand enemies that threaten them; for, as long as the frosts of winter have not arrived, these open galleries are trodden by Spiders, by Acari, by Anthrenus-grubs and other plunderers, to whom these eggs, or the young larvæ about to emerge from them, must be a dainty feast. In consequence of the mother's heedlessness, the number of those who escape all these voracious hunters and the inclemencies of the weather must be curiously small. This perhaps explains why she is compelled to make up by her fecundity for her deficient industry.

The hatching occurs a month later, about the end of September or the beginning of October. The season being still propitious, I was led to suppose that the young larvæ must at once make a start and disperse, in order that each might seek to gain access, through some imperceptible fissure, to an Anthophora-cell. This presumption turned out to be entirely at fault. In the boxes in which I had placed the eggs laid by my captives, the young larvæ, little black creatures at most a twenty-fifth of an inch long, did not move away, provided though they were with vigorous legs; they remained higgledy-piggledy with the white skins of the eggs whence they had emerged.

In vain I placed within their reach lumps of earth containing nests of the Anthophora, open cells, larvæ and nymphs of the Bee: nothing was able to tempt them; they persisted in forming, with the egg-skins, a powdery heap of speckled black and white. It was only by drawing the point of a needle through this pinch of living dust that I was able to provoke an active wriggling. Apart from this, all was still. If I forcibly removed a few larvæ from the common heap, they at once hurried back to it, in order to hide themselves among the rest. Perhaps they had less reason to fear the cold when thus collected and sheltered beneath the egg-skins. Whatever may be the motive that impels them to remain thus gathered in a heap, I recognized that none of the means suggested by my imagination succeeded in forcing them to abandon the little spongy mass formed by the skins of the eggs, which were slightly glued together. Lastly, to assure myself that the larvæ, in the free state, do not disperse after they are hatched, I went during the winter to Carpentras and inspected the banks inhabited by the Anthophoræ. There, as in my boxes, I found the larvæ piled into heaps, all mixed up with the skins of the eggs.

CHAPTER III
THE PRIMARY LARVA OF THE SITARES

Nothing new happens before the end of the following April. I shall profit by this long period of repose to tell you more about the young larva, of which I will begin by giving a description. Its length is a twenty-fifth of an inch, or a little less. It is hard as leather, a glossy greenish black, convex above and flat below, long and slender, with a diameter increasing gradually from the head to the hinder extremity of the metathorax, after which it rapidly diminishes. Its head is a trifle longer than it is wide and is slightly dilated at the base; it is pale-red near the mouth and darker about the ocelli.

The labrum forms a segment of a circle; it is reddish, edged with a small number of very short, stiff hairs. The mandibles are powerful, red-brown, curved and sharp; when at rest they meet without crossing. The maxillary palpi are rather long, consisting of two cylindrical sections of equal length, the outer ending in a very short bristle. The jaws and the lower lip are not sufficiently visible to lend themselves to accurate description.

The antennæ consist of two cylindrical segments, equal in length, not very definitely divided; these segments are nearly as long as those of the palpi; the outer is surmounted by a cirrus whose length is as much as thrice that of the head and tapers off until it becomes invisible under a powerful pocket-lens. Behind the base of either antennæ are two ocelli, unequal in size and almost touching.

The thoracic segments are of equal length and increase gradually in width from front to back. The prothorax is wider than the head, but is narrower in front than at the base and is slightly rounded at the sides. The legs are of medium length and fairly robust, ending in a long, powerful, sharp and very mobile claw. On the haunch and thigh of each leg is a long cirrus, like that of the antennæ, almost as long as the whole limb and standing at right angles to the plane of locomotion when the creature moves. There are a few stiff bristles on the legs.

The abdomen has nine segments, of practically equal length, but shorter than those of the thorax and diminishing very rapidly in width toward the last. Fixed below the eighth segment, or rather below the strip of membrane separating this segment and the last, we see two spikes, slightly curved, short, but with strong, sharp, hard points, and placed one to the right and the other to the left of the median line. These two appendages are able, by means of a mechanism recalling, on a smaller scale, that of the Snail's horns, to withdraw into themselves, as a result of the membranous character of their base. They can also retreat under the eighth segment, borne, as they are, by the anal segment, when this last, as it contracts, withdraws into the eighth. Lastly, the ninth or anal segment bears on its hinder edge two long cirri, like those of the legs and the antennæ, curving backwards from tip to base. At the rear of this segment a fleshy nipple appears, more or less prominent; this is the anus. I do not know where the stigmata are placed; they have evaded my investigations, though these were undertaken with the aid of the microscope.

When the larva is at rest, the various segments overlap evenly; and the membranous intervals, corresponding with the articulations, do not show. But, when the larva walks, all the articulations, especially those of the abdominal segments, are distended and end by occupying almost as much space as the horny arches. At the same time the anal segment emerges from the sheath formed by the eighth; the anus, in turn, is stretched into a nipple; and the two points of the penultimate ring rise, at first slowly, and then suddenly stand up with an abrupt motion similar to that of a spring when released. In the end, these two points diverge like the horns of a crescent. Once this complex apparatus is unfolded, the tiny creature is ready to crawl upon the most slippery surface.

The last segment and its anal button are curved at right angles to the axis of the body; and the anus comes and presses upon the surface of locomotion, where it ejects a tiny drop of transparent, treacly fluid, which glues and holds the little creature firmly in position, supported on a sort of tripod formed by the anal button and the two cirri of the last segment. If we are observing the animal's manner of locomotion on a strip of glass, we can hold the strip in a vertical position, or even turn it upside down, or shake it lightly, without causing the larva to become detached and fall, held fast as it is by the glutinous secretion of the anal button.

If it has to proceed along a surface where there is no danger of a fall, the microscopic creature employs another method. It crooks its belly and, when the two spikes of the eighth segment, now fully outspread, have found a firm support by ploughing, so to speak, the surface of locomotion, it bears upon that base and pushes forward by expanding the various abdominal articulations. This forward movement is also assisted by the action of the legs, which are far from remaining inactive. This done, it casts anchor with the powerful claws of its feet; the abdomen contracts; the various segments draw together; and the anus, pulled forward, obtains a fresh purchase, with the aid of the two spikes, before beginning the second of these curious strides.

During these manoeuvres, the cirri of the flanks and thighs drag along the supporting surface and by their length and elasticity appear fitted only to impede progress. But let us not be in a hurry to conclude that we have discovered an inconsistency: the least of creatures is adapted to the conditions amid which it has to live; there is reason to believe that these filaments, far from hampering the pigmy's progress, must, in normal circumstances, be of some assistance to it.

Even the little that we have just learnt shows us that the young Sitaris-larva is not called upon to move on an ordinary surface. The spot, whatever it may be like, where this larva is to live later exposes it to the risk of many dangerous falls, since, in order to prevent them, it is not only equipped with strong and extremely mobile talons and a steel-shod crescent, a sort of ploughshare capable of biting into the most highly polished substance, but is further provided with a viscous liquid, sufficiently tenacious and adhesive to hold it in position without the help of other appliances. In vain I racked my brains to guess what the substance might be, so shifting, so uncertain and so perilous, which the young Sitares are destined to inhabit; and

I discovered nothing to explain the necessity for the structure which I have described. Convinced beforehand, by an attentive examination of this structure, that I should witness some peculiar habits, I waited with eager impatience for the return of the warm weather, never doubting that by dint of persevering observation the mystery would be disclosed to me next spring. At last this spring, so fervently desired, arrived; I brought to bear all the patience, all the imagination, all the insight and discernment that I may possess; but, to my utter shame and still greater regret, the secret escaped me. Oh, how painful are those tortures of indecision, when one has to postpone till the following year an investigation which has led to no result!

My observations made during the spring of 1856, although purely negative, nevertheless have an interest of their own, because they prove the inaccuracy of certain suppositions to which the undeniable parasitism of the Sitares naturally inclines us. I will therefore relate them in a few words. At the end of April, the young larvæ, hitherto motionless and concealed in the spongy heap of the egg-skins, emerge from their immobility, scatter and run about in all directions through the boxes and jars in which they have passed the winter. By their hurried gait and their indefatigable evolutions we readily guess that they are seeking something which they lack. What can this something be, unless it be food? For remember that these larvæ were hatched at the end of September and that since then, that is to say, for seven long months, they have taken no nourishment, though they have spent this period in the full enjoyment of their vitality, as I was able to assure myself all through the winter by irritating them, and not in a state of torpor similar to that of the hibernating animals. From the moment of their hatching they are doomed, although full of life, to an absolute abstinence of seven months' duration; and it is natural to suppose, when we see their present excitement, that an imperious hunger sets them bustling in this fashion.

The desired nourishment could only be the contents of the cells of the Anthophora, since we afterwards find the Sitares in these cells. Now these contents are limited to honey or larvæ. It just happens that I have kept some Anthophora-cells occupied by larvæ or nymphs. I place a few of these, some open, some closed, within reach of the young Sitares, as I had already done directly after the hatching. I even slip the Sitares into the cells: I place them on the sides of the larva, a succulent morsel to all appearances; I do all sorts of things to tempt their appetite; and, after exhausting my ingenuity, which continues fruitless, I remain convinced that my famished grubs are seeking neither the larvæ nor nymphs of the Anthophora.

Let us now try honey. We must obviously employ honey prepared by the same species of Anthophora as that at whose cost the Sitares live. But this Bee is not very common in the neighbourhood of Avignon; and my engagements at the college[1] do not allow me to absent myself for the purpose of repairing to Carpentras, where she is so abundant. In hunting for cells provisioned with honey I thus lose a good part of the month of May; however, I end by finding some which are newly sealed and which belong to the right Anthophora. I open these cells with the feverish impatience of a sorely-tried longing. All goes well: they are half-full of fluid, dark, nauseating honey, with the Bee's lately-hatched larva floating on the surface. This larva is removed; and taking a thousand precautions, I lay one or more Sitares on the surface of the honey. In other cells I leave the Bee's larva and insert Sitares, placing them sometimes on the honey and sometimes on the inner wall of the cell or simply at the entrance. Lastly, all the cells thus prepared are put in glass tubes, which enable me to observe them readily, without fear of disturbing my famished guests at their meal.

[1] Fabre, as a young man, was a master at Avignon College. Cf. *The Life of the Fly:* chaps. xii., xiii., xix. and xx.—*Translator's Note.*

But what am I saying? Their meal? There is no meal! The Sitares, placed at the entrance to a cell, far from seeking to make their way in, leave it and go roaming about the glass tube; those which have been placed on the inner surface of the cells, near the honey, emerge precipitately, half-caught in the glue and tripping at every step; lastly, those which I thought I had favoured the most, by placing them on the honey itself, struggle, become entangled in the sticky mass and perish in it, suffocated. Never did experiment break down so completely! Larvæ, nymphs, cells, honey: I have offered you them all! Then what do you want, you fiendish little creatures?

Tired of all these fruitless attempts, I ended where I ought to have begun: I went to Carpentras. But it was too late: the Anthophora had finished her work; and I did not succeed in seeing anything new. During the course of the year I learnt from Léon Dufour,[2] to whom I had spoken of the Sitares, that the tiny creature which he had found on the Andrenæ[3] and described under the generic name of Triungulinus, was recognized later by Newport[4] as the larva of a Meloe, or Oil-beetle. Now it so happened that I had found a few Oil-beetles in the cells of the same Anthophora that nourishes the Sitares. Could there be a similarity of habits between the two kinds of insects? This idea threw a sudden light for me upon the subject; but I had plenty of time in which to mature my plans: I had another year to wait.

[2] Jean Marie Léon Dufour (1780-1865), an army surgeon who served with distinction in several campaigns, and subsequently practised as a doctor in the Landes, where he attained great eminence as a naturalist. Fabre often refers to him as the Wizard of the Landes. Cf. *The Life of the Spider*, by J. Henri Fabre, translated by Alexander Teixeira de Mattos: chap. i.; and *The Life of the Fly:* chap. i.—*Translator's Note.*

[3] A genus of Burrowing Bee, the most numerous in species among the British Bees.—*Translator's Note.*

[4] George Newport (1803-1854), an English surgeon and naturalist, president of the Entomological Society from 1844 to 1845 and an expert in insect anatomy.—*Translator's Note.*

When April came, my Sitaris-larvæ began, as usual, to bestir themselves. The first Bee to appear, an Osmia, is dropped alive into a glass jar containing a few of these larvæ; and after a lapse of some fifteen minutes I inspect them through the pocket-lens. Five Sitares are embedded in the fleece of the thorax. It is done, the problem's solved! The larvæ of the Sitares, like those of the Oil-beetles, cling like grim death to the fleece of their generous host and make him carry them into the cell. Ten times over I repeat the experiment with the various Bees that come to plunder the lilac flowering outside my window and in particular with male Anthophoræ; the result is still the same: the larvæ embed themselves in the hair of the Bees' thorax. But after so many disappointments one becomes distrustful and it is better to go and observe the facts upon the spot; besides, the Easter holidays fall very conveniently and afford me the leisure for my observations.

I will admit that my heart was beating a little faster than usual when I found myself once again standing in front of the perpendicular bank in which the Anthophora nests. What will be the result of the experiment? Will it once more cover me with confusion? The weather is cold and rainy; not a Bee shows herself on the few spring flowers that have come out. Numbers of Anthophoræ cower, numbed and motionless, at the entrance to the galleries. With the tweezers, I extract them one by one from their lurking-places, to examine them under the lens. The first has Sitaris-larvæ on her thorax; so has the second; the third and fourth likewise; and so on, as far as I care to pursue the examination. I change galleries ten times, twenty times; the result is invariable. Then, for me, occurred one of the moments which come to those who, after considering and reconsidering an idea for years and years from every point of view, are at last able to cry: "Eureka!"

On the days that followed, a serene and balmy sky enabled the Anthophoræ to leave their retreats and scatter over the countryside and despoil the flowers. I renewed my examination on those Anthophoræ flying incessantly from one flower to another, whether in the neighbourhood of the places where they were born or at great distances from these places. Some were without Sitaris-larvæ; others, more numerous, had two, three, four, five or more among the hairs of their thorax. At Avignon, where I have not yet seen *Sitaris humeralis*, the same species of Anthophora, observed at almost the same season, while pillaging the lilac-blossom, was always free of young Sitaris-grubs; at Carpentras, on the contrary, where there is not a single Anthophora-colony without Sitares, nearly three-quarters of the specimens which I examined carried a few of these larvæ in their fleece.

But, on the other hand, if we look for these larvæ in the entrance-lobbies where we found them, a few days ago, piled up in heaps, we no longer see them. Consequently, when the Anthophoræ, having opened their cells, enter the galleries to reach the exit and fly away, or else when the bad weather and the darkness bring them back there for a time, the young Sitaris-larvæ, kept on the alert in these same galleries by the stimulus of instinct, attach themselves to the Bees, wriggling into their fur and clutching it so firmly that they need not fear a fall during the long journeys of the insect which carries them. By thus attaching themselves to the Anthophoræ the young Sitares evidently intend to get themselves carried, at the opportune moment, into the victualled cells.

One might even at first sight believe that they live for some time on the Anthophora's body, just as the ordinary parasites, the various species of Lice, live on the body of the animal that feeds them. But not at all. The young Sitares, embedded in the fleece, at right angles to the Anthophora's body, head inwards, rump outwards, do not stir from the point which they have selected, a point near the Bee's shoulders. We do not see them wandering from spot to spot, exploring the Anthophora's body, seeking the part where the skin is more delicate, as they would certainly do if they were really deriving some nourishment from the juices of the Bee. On the contrary, they are nearly always established on the toughest and hardest part of the Bee's body, on the thorax, a little below the insertion of the wings, or, more rarely, on the head; and they remain absolutely motionless, fixed to the same hair, by means of the mandibles, the feet, the closed crescent of the eighth segment and, lastly, the glue of the anal button. If they chance to be disturbed in this position, they reluctantly repair to another point of the thorax, pushing their way through the insect's fur and in the end fastening on to another hair, as before.

To confirm my conviction that the young Sitaris-grubs do not feed on the Anthophora's body, I have sometimes placed within their reach, in a glass jar, some Bees that have long been dead and are completely dried up. On these dry corpses, fit at most for gnawing, but certainly containing nothing to suck, the Sitaris-larvæ took up their customary position and there remained motionless as on the living insect. They obtain nothing, therefore, from the Anthophora's body; but perhaps they nibble her fleece, even as the Bird-lice nibble the birds' feathers?

To do this, they would require mouth-parts endowed with a certain strength and, in particular, horny and sturdy jaws, whereas their jaws are so fine that a microscopic examination failed to show them to me. The larvæ, it is true, are provided with powerful mandibles; but these finely-pointed mandibles, with their backward curve, though excellent for tugging at food and tearing it to pieces, are useless for grinding it or gnawing it. Lastly, we have a final proof of the passive condition of the Sitaris-larvæ on the body of the Anthophoræ in the fact that the Bees do not appear to be in any way incommoded by their presence, since we do not see them trying to rid themselves of the grubs. Some Anthophoræ which were free from these grubs and some others which were carrying five or six upon their bodies were placed separately in glass jars. When the first disturbance resulting from their captivity was appeased, I could see nothing peculiar about those occupied by the young Sitares. And, if all these arguments were not sufficient, I might add that a creature which has already been able to spend seven months without food and which in a few days' time will proceed to drink a highly-flavoured fluid would be guilty of a singular inconsistency if it were to start nibbling the dry fleece of a Bee. It therefore seems to me undeniable that the young Sitares settle on the Anthophora's body merely to make her carry them into the cells which she will soon be building.

But until then the future parasites must hold tight to the fleece of their hostess, despite her rapid evolutions among the flowers, despite her rubbing against the walls of the galleries when she enters to take shelter and, above all, despite the brushing which she must often give herself with her feet to dust herself and keep spick and span. Hence no doubt the need for that curious apparatus which no standing or moving upon ordinary surfaces could explain, as was said above, when we were wondering what the shifting, swaying, dangerous body might be on which the larva would have to establish itself later. This body is a hair of a Bee who makes a thousand rapid journeys, now diving into her narrow galleries, now forcing her way down the tight throat of a corolla, and who never rests except to brush herself with her feet and remove the specks of dust collected by the down which covers her.

We can now easily understand the use of the projecting crescent whose two horns, by closing together, are able to take hold of a hair more easily than the most delicate tweezers; we perceive the full value of the tenacious adhesive provided by the anus to save the tiny creature, at the least sign of danger, from an imminent fall; we realize lastly the useful function that may be fulfilled by the elastic cirri of the flanks and legs, which are an absolute and most embarrassing superfluity when walking upon a smooth surface, but which, in the present case, penetrate like so many probes into the thickness of the Anthophora's down and serve as it were to anchor the Sitaris-larva in position. The more we consider this arrangement, which seems modelled by a blind caprice so long as the grub drags itself laboriously over a smooth surface, the more do we marvel at the means, as effective as they are varied, which are lavished upon this fragile creature to help it to preserve its unstable equilibrium.

Before I describe what becomes of the Sitaris-grubs on leaving the body of the Anthophoræ, I must not omit to mention one very remarkable peculiarity. All the Bees invaded by these grubs that have hitherto been observed have, without one exception, been male Anthophoræ. Those whom I drew from their lurking-places were males; those whom I caught upon the flowers were males; and, in spite of the most active search, I failed to find a single female at liberty. The cause of this total absence of females is easy to understand.

If we remove a few clods from the area occupied by the nests, we see that, though all the males have already opened and abandoned their cells, the females, on the contrary, are still enclosed in theirs, but on the point of soon taking flight. This appearance of the males almost a month before the emergence of the females is not peculiar to the Anthophoræ; I have observed it in many other Bees and particularly in the Three-horned Osmia (*O. tricornis*), who inhabits the same site as the Hairy-footed Anthophora (*A. pilipes*). The males of the Osmia make their appearance even before those of the Anthophora and at so early a season that the young Sitaris-larvæ are perhaps not yet aroused by the instinctive impulse which urges them to activity. It is no doubt to their precocious awakening that the males of the Osmia owe their ability to traverse with impunity the corridors in which the young Sitaris-grubs are heaped together, without having the latter fasten to their fleece; at least, I cannot otherwise explain the absence of these larvæ from the backs of the male Osmiæ, since, when we place them artificially in the presence of these Bees, they fasten on them as readily as on the Anthophoræ.

The emergence from the common site begun by the male Osmiæ is continued by the male Anthophoræ and ends with the almost simultaneous emergence of the female Osmiæ and Anthophoræ. I was easily able to verify this sequence by observing at my own place, in the early spring, the dates at which the cells, collected during the previous autumn, were broken.

At the moment of their emergence, the male Anthophoræ, passing through the galleries in which the Sitaris-larvæ are waiting on the alert, must pick up a certain number of them; and those among them who, by entering empty corridors, escape the enemy on this first occasion will not evade him for long, for the rain, the chilly air and the darkness bring them back to their former homes, where they take shelter now in one gallery, now in another, during a great part of April. This constant traffic of the males in the entrance-lobbies of their houses and the prolonged stay which the bad weather often compels them to make provide the Sitares with the most favourable opportunity for slipping into the Bees' fur and taking up their position. Moreover, when this state of affairs has lasted a month or so, there can be only very few if any larvæ left wandering about without having attained their end. At that period I was unable to find them anywhere save on the body of the male Anthophora.

It is therefore extremely probable that, on their emergence, which takes place as May draws near, the female Anthophoræ do not pick up Sitaris-larvæ in the corridors, or pick up only a number which will not compare with that carried by the males. In fact, the first females that I was able to observe in April, in the actual neighbourhood of the nests, were free from these larvæ. Nevertheless it is upon the females that the Sitaris-larvæ must finally establish themselves, for the males upon whom they now are cannot introduce them into the cells, since they take no part in the building or provisioning. There is therefore, at a given moment, a transfer of Sitaris-larvæ from the male Anthophoræ to the females; and this transfer is, beyond a doubt, effected during the union of the sexes. The female finds in the male's embraces both life and death for her offspring; at the moment when she surrenders herself to the male for the preservation of her race, the vigilant parasites pass from the male to the female, with the extermination of that same race in view.

In support of these deductions, here is a fairly conclusive experiment, though it reproduces the natural circumstances but roughly. On a female taken in her cell and therefore free from Sitares, I place a male who is infested with them; and I keep the two sexes in contact, suppressing their unruly movements as far as I am able. After fifteen or twenty minutes of this enforced proximity, the female is invaded by one or more of the larvæ which at first were on the male. True, experiment does not always succeed under these imperfect conditions.

By watching at Avignon the few Anthophoræ that I succeeded in discovering, I was able to detect the precise moment of their work; and on the following Thursday,[5] the 21st of May, I repaired in all haste to Carpentras, to witness, if possible, the entrance of the Sitares into the Bee's cells. I was not mistaken: the works were in full swing.

[5] Thursday is the weekly holiday in French schools.—*Translator's Note*.

In front of a high expanse of earth, a swarm stimulated by the sun, which floods it with light and heat, is dancing a crazy ballet. It is a hover of Anthophoræ, a few feet thick and covering an area which matches the sort of house-front formed by the perpendicular soil. From the tumultuous heart of the cloud rises a monotonous, threatening murmur, while the bewildered eye strays through the inextricable evolutions of the eager throng. With the rapidity of a lightning-flash thousands of Anthophoræ are incessantly flying off and scattering over the country-side in search of booty; thousands of others also are incessantly arriving, laden with honey or mortar, and keeping up the formidable proportions of the swarm.

I was at that time something of a novice as regards the nature of these insects:

"Woe," said I to myself, "woe to the reckless wight bold enough to enter the heart of this swarm and, above all, to lay a rash hand upon the dwellings under construction! Forthwith surrounded by the furious host, he would expiate his rash attempt, stabbed by a thousand stings!"

At this thought, rendered still more alarming by the recollection of certain misadventures of which I had been the victim when seeking to observe too closely the combs of the Hornet (*Vespa crabro*), I felt a shiver of apprehension pass through my body.

Yet, to obtain light upon the question which brings me hither, I must needs penetrate the fearsome swarm; I must stand for whole hours, perhaps all day, watching the works which I intend to upset; lens in hand, I must scrutinize, unmoved amid the whirl, the things that are happening in the cells. The use moreover of a mask, of gloves, of a covering of any kind is impracticable, for utter dexterity of the fingers and complete liberty of sight are essential to the investigations which I have to make. No matter: even though I leave this wasps'-nest with a face swollen beyond recognition, I must to-day obtain a decisive solution of the problem which has preoccupied me too long.

A few strokes of the net, aimed, beyond the limits of the swarm, at the Anthophoræ on their way to the harvest or returning, soon informed me that the Sitaris-larvæ are perched on the thorax, as I expected, occupying the same position as on the males. The circumstances therefore could not be more favourable. We will inspect the cells without further delay.

My preparations are made at once: I button my clothes tightly, so as to afford the Bees the least possible opportunity, and I enter the heart of the swarm. A few blows of the mattock, which arouse a far from reassuring crescendo in the humming of the Anthophoræ, soon place me in possession of a lump of earth; and I beat a hasty retreat, greatly astonished to find myself still safe and sound and unpursued. But the lump of earth which I have removed is from a part too near the surface; it contains nothing but Osmia-cells, which do not interest me for the moment. A second expedition is made, lasting longer than the first; and, though my retreat is effected without great precipitation, not an Anthophora has touched me with her sting, nor even shown herself disposed to fall upon the aggressor.

This success emboldens me. I remain permanently in front of the work in progress, continually removing lumps of earth filled with cells, spilling the liquid honey on the ground, eviscerating larvæ and crushing the Bees busily occupied in their nests. All this devastation results merely in arousing a louder hum in the swarm and is not followed by any hostile demonstration. The Anthophoræ whose cells are not hurt go about their labours as if nothing unusual were happening round about them; those whose dwellings are overturned try to repair them, or hover distractedly in front of the ruins; but none of them seems inclined to swoop down upon the author of the damage. At most, a few, more irritated than the rest, come at intervals and hover before my face, confronting me at a distance of a couple of inches, and then fly away, after a few moments of this curious inspection.

Despite the selection of a common site for their nests, which might suggest an attempt at communistic interests among the Anthophoræ, these Bees, therefore, obey the egotistical law of each one for himself and do not know how to band themselves together to repel an

The Glow-Worm and Other Beetles

enemy who threatens one and all. Taken singly, the Anthophora does not even know how to dash at the enemy who is ravaging her cells and drive him away with her stings; the pacific creature hastily leaves its dwelling when disturbed by undermining and escapes in a crippled state, sometimes even mortally wounded, without thinking of making use of its venomous sting, except when it is seized and handled. Many other Hymenoptera, honey-gatherers or hunters, are quite as spiritless; and I can assert to-day, after a long experience, that only the Social Hymenoptera, the Hive-bees, the Common Wasps and the Bumble-bees, know how to devise a common defence; and only they dare fall singly upon the aggressor, to wreak an individual vengeance.

Thanks to this unexpected lack of spirit in the Mason-bee, I was able for hours to pursue my investigations at my leisure, seated on a stone in the midst of the murmuring and distracted swarm, without receiving a single sting, though I took no precautions whatever. Country-folk, happening to pass and beholding me seated, unperturbed, in the midst of the whirl of Bees, stopped aghast to ask me whether I had bewitched them, whether I charmed them, since I appeared to have nothing to fear from them:

"Mé, moun bel ami, li-z-avé doun escounjurado què vous pougnioun pas, canèu de sort!"

My miscellaneous impedimenta spread over the ground, boxes, glass jars and tubes, tweezers and magnifying-glasses, were certainly regarded by these good people as the implements of my wizardry.

We will now proceed to examine the cells. Some are still open and contain only a more or less complete store of honey. Others are hermetically sealed with an earthen lid. The contents of these latter vary greatly. Sometimes we find the larva of a Bee which has finished its mess or is on the point of finishing it; sometimes a larva, white like the first, but more corpulent and of a different shape; at other times honey with an egg floating on the surface. The honey is liquid and sticky, with a brownish colour and a very strong, repulsive smell. The egg is of a beautiful white, cylindrical in shape, slightly curved into an arc, a fifth or a sixth of an inch in length and not quite a twenty-fifth of an inch in thickness; it is the egg of the Anthophora.

In a few cells this egg is floating all alone on the surface of the honey; in others, very numerous these, we see, lying on the egg of the Anthophora, as on a sort of raft, a young Sitaris-grub with the shape and the dimensions which I have described above, that is to say, with the shape and the dimensions which the creature possesses on leaving the egg. This is the enemy within the gates.

When and how did it get in? In none of the cells where I have observed it was I able to distinguish a fissure which could have allowed it to enter; they are all sealed in a quite irreproachable manner. The parasite therefore established itself in the honey-warehouse before the warehouse was closed; on the other hand, the open cells, full of honey, but as yet without the egg of the Anthophora, are always free from parasites. It is therefore during the laying, or afterwards, when the Anthophora is occupied in plastering the door of the cell, that the young larva gains admittance. It is impossible to decide by experiment to which of these two periods we must ascribe the introduction of the Sitares into the cell; for, however peaceable the Anthophora may be, it is evident that we cannot hope to witness what happens in the cell at the moment when she is laying an egg or at the moment when she is making the lid. But a few attempts will soon convince us that the only second which would allow the Sitaris to establish itself in the home of the Bee is the very second when the egg is laid on the surface of the honey.

Let us take an Anthophora-cell full of honey and furnished with an egg and, after removing the lid, place it in a glass tube with a few Sitaris-grubs. The grubs do not appear at all eager for this wealth of nectar placed within their reach; they wander at random about the tube, run about the outside of the cell, sometimes happen upon the edge of the orifice and very rarely venture inside. When they do, they do not go far in and they come out again at once. If one happens to reach the honey, which only half fills the cell, it tries to escape as soon as it has perceived the shifting nature of the sticky soil upon which it was about to enter; but, tottering at every step, because of the viscous matter clinging to its feet, it often ends by falling back into the honey, where it dies of suffocation.

Again, we may experiment as follows: having prepared a cell as before, we place a larva most carefully on its inner wall, or else on the surface of the food itself. In the first case, the larva hastens to leave the cell; in the second case, it struggles awhile on the surface of the honey and ends by getting so completely caught that, after a thousand efforts to gain the shore, it is swallowed up in the viscous lake.

In short, all attempts to establish the Sitaris-grub in an Anthophora-cell provisioned with honey and furnished with an egg are no more successful than those which I made with cells whose store of food had already been broached by the larva of the Bee, as described above. It is therefore certain that the Sitaris-grub does not leave the fleece of the Mason-bee when the Bee is in her cell or at the entrance to it, in order itself to make a rush for the coveted honey; for this honey would inevitably cause its death, if it happened by accident to touch the perilous surface merely with the tip of its tarsi.

Since we cannot admit that the Sitaris-grub leaves the furry corselet of its hostess to slip unseen into the cell, whose orifice is not yet wholly walled up, at the moment when the Anthophora is building her door, all that remains to investigate is the second at which the egg is being laid. Remember in the first place that the young Sitaris which we find in a closed cell is always placed on the egg of the Bee. We shall see in a minute that this egg not merely serves as a raft for the tiny creature floating on a very treacherous lake, but also constitutes the first and indispensable part of its diet. To get at this egg, situated in the centre of the lake of honey, to reach, at all costs, this raft, which is also its first ration, the young larva evidently possesses some means of avoiding the fatal contact of the honey; and this means can be provided only by the actions of the Bee herself.

In the second place, observations repeated *ad nauseam* have shown me that at no period do we find in each invaded cell more than a single Sitaris, in one or other of the forms which it successively assumes. Yet there are several young larvæ established in the silky tangle of the Bee's thorax, all eagerly watching for the propitious moment at which to enter the dwelling in which they are to continue their development. How then does it happen that these larvæ, goaded by such an appetite as one would expect after seven or eight months' complete abstinence, instead of all rushing together into the first cell within reach, on the contrary enter the various cells which the Bee is provisioning one at a time and in perfect order? Some action must take place here independent of the Sitares.

To satisfy those two indispensable conditions, the arrival of the larva upon the egg without crossing the honey and the introduction of a single larva among all those waiting in the fleece of the Bee, there can be only one explanation, which is to suppose that, at the moment when the Anthophora's egg is half out of the oviduct, one of the Sitares which have hastened from the thorax to the tip of the abdomen, one more highly favoured by its position, instantly settles upon the egg, a bridge too narrow for two, and with it reaches the surface of the honey. The impossibility of otherwise fulfilling the two conditions which I have stated gives to the explanation which I am offering a degree of certainty almost equivalent to that which would be furnished by direct observation, which is here, unfortunately, impracticable. This presupposes, it is true, in the microscopic little creature destined to live in a place where so many dangers threaten it from the first, an astonishingly rational inspiration, which adapts the means to the end with amazing logic. But is not this the invariable conclusion to which the study of instinct always leads us?

When dropping her egg upon the honey, therefore, the Anthophora at the same time deposits in her cell the mortal enemy of her race; she carefully plasters the lid which closes the entrance to the cell; and all is done. A second cell is built beside it, probably to suffer the same fatal doom; and so on until the more or less numerous parasites sheltered by her down are all accommodated. Let us leave the unhappy mother to continue her fruitless task and turn our attention to the young larva which has so adroitly secured itself board and lodging.

In opening cells whose lid is still moist, we end by discovering one in which the egg, recently laid, supports a young Sitaris. This egg is intact and in irreproachable condition. But now the work of devastation begins: the larva, a tiny black speck which we see running over the white surface of the egg, at last stops and balances itself firmly on its six legs; then, seizing the delicate skin of the egg with the sharp hooks of its mandibles, it tugs at it violently until it breaks, spilling its contents, which the larva eagerly drinks up. Thus the first stroke of the mandibles which the parasite delivers in the usurped cell is aimed at the destruction of the Bee's egg. A highly logical precaution! The Sitaris-larva, as we shall see, has to feed upon the honey in the cell; the Anthophora-larva which would proceed from that egg would require the same food; but the portion is too small for two; so, quick, a bite at the egg and the difficulty will be removed. The story of these facts calls for no comment. This destruction of the cumbersome egg is all the more inevitable inasmuch as special tastes compel the young Sitaris-grub to make its first meals of it. Indeed we see the tiny creature begin by greedily drinking the juices which the torn wrapper of the egg allows to escape; and for several days it may be observed, at one time motionless on this envelope, in which it rummages at intervals with its head, at others running over it from end to end to rip it open still wider and to cause a little of the juices, which become daily less abundant, to trickle from it; but we never catch it imbibing the honey which surrounds it on every side.

For that matter, it is easy to convince ourselves that the egg combines with the function of a life-buoy that of the first ration. I have laid on the surface of the honey in a cell a tiny strip of paper, of the same dimensions as the egg; and on this raft I have placed a Sitaris-larva. Despite every care, my attempts, many times repeated, always failed. The larva, placed in a paper boat in the centre of the mass of honey, behaves as in the earlier experiments. Not finding what suits it, it tries to escape and perishes in the sticky toils as soon as it leaves the strip of paper, which it soon does.

On the other hand, we can easily rear Sitaris-grubs by taking Anthophora-cells not invaded by the parasites, cells in which the egg is not yet hatched. All that we have to do is to pick up one of these grubs with the moistened tip of a needle and to lay it delicately on the egg. There is then no longer the least attempt to escape. After exploring the egg to find its way about, the larva rips it open and for several days does not stir from the spot. Henceforth its development takes place unhindered, provided that the cell be protected from too rapid evaporation, which would dry up the honey and render it unfit for the grub's food. The Anthophora's egg therefore is absolutely necessary to the Sitaris-larva, not merely as a boat, but also as its first nourishment. This is the whole secret, for lack of knowing which I had hitherto failed in my attempts to rear the larvæ hatched in my glass jars.

At the end of a week, the egg, drained by the parasite, is nothing but a dry skin. The first meal is finished. The Sitaris-larva, whose dimensions have almost doubled, now splits open along the back; and through a slit which comprises the head and the three thoracic segments a white corpusculum, the second form of this singular organism, escapes to fall on the surface of the honey, while the abandoned slough remains clinging to the raft which has hitherto safeguarded and fed the larva. Presently both sloughs, those of the Sitaris and the egg, will disappear, submerged under the waves of honey which the new larva is about to raise. Here ends the history of the first form adopted by the Sitaris.

In summing up the above, we see that the strange little creature awaits, without food, for seven months, the appearance of the Anthophoræ and at last fastens on to the hairs on the corselet of the males, who are the first to emerge and who inevitably pass within its reach in going through their corridors. From the fleece of the male the larva moves, three or four weeks later, to that of the female, at the moment of coupling; and then from the female to the egg leaving the oviduct. It is by this concatenation of complex manoeuvres that the larva in the end finds itself perched upon an egg in the middle of a closed cell filled with honey. These perilous gymnastics on the hair of a Bee in movement all the day, this passing from one sex to the other, this arrival in the middle of the cell by way of the egg, a dangerous bridge thrown across the sticky abyss, all this necessitates the balancing-appliances with which it is provided and which I have described above. Lastly, the destruction of the egg calls, in its turn, for a sharp pair of scissors; and such is the object of the keen, curved mandibles. Thus the primary form of the Sitares has as its function to get itself carried by the Anthophora into the cell and to rip up her egg. This done, the organism becomes transformed to such a degree that repeated observations are required to make us believe the evidence of our eyes.

CHAPTER IV
THE PRIMARY LARVA OF THE OIL-BEETLES

I interrupt the history of the Sitares to speak of the Meloes, those uncouth Beetles, with their clumsy belly and their limp wing-cases yawning over their back like the tails of a fat man's coat that is far too tight for its wearer. The insect is ugly in colouring, which is black, with an occasional blue gleam, and uglier still in shape and gait; and its disgusting method of defence increases the repugnance with which it inspires us. If it judges itself to be in danger, the Meloe resorts to spontaneous bleeding. From its joints a yellowish, oily fluid oozes, which stains your fingers and makes them stink. This is the creature's blood. The English, because of its trick of discharging oily blood when on the defensive, call this insect the Oil-beetle. It would not be a particularly interesting Beetle save for its metamorphoses and the peregrinations of its larva, which are similar in every respect to those of the larva of the Sitares. In their first form, the Oil-beetles are parasites of the Anthophoræ; their tiny grub, when it leaves the egg, has itself carried into the cell by the Bee whose victuals are to form its food.

Observed in the down of various Bees, the queer little creature for a long time baffled the sagacity of the naturalists, who, mistaking its true origin, made it a species of a special family of wingless insects. It was the Bee-louse (*Pediculus apis*) of Linnæus;[1] the Triungulin of the Andrenæ (*Triungulinus andrenetarum*) of Léon Dufour. They saw in it a parasite, a sort of Louse, living in the fleece of the honey-gatherers. It was reserved for the distinguished English naturalist Newport to show that this supposed Louse was the first state of the Oil-beetles. Some observations of my own will fill a few lacunæ in the English scientist's monograph. I will therefore sketch the evolution of the Oil-beetles, using Newport's work where my own observations are defective. In this way the Sitares and the Meloes, alike in habits and transformations, will be compared; and the comparison will throw a certain light upon the strange metamorphoses of these insects.

[1] Carolus Linnæus (Karl von Linné, 1707-1778), the celebrated Swedish botanist and naturalist, founder of the Linnæan system of classification.—*Translator's Note.*

The same Mason-bee (*Anthophora pilipes*) upon whom the Sitares live also feeds a few scarce Meloes (*M. cicatricosus*) in its cells. A second Anthophora of my district (*A. parietina*) is more subject to this parasite's invasions. It was also in the nests of an Anthophora, but of a different species (*A. retusa*), that Newport observed the same Oil-beetle. These three lodgings adopted by *Meloe cicatricosus* may be of some slight interest, as leading us to suspect that each species of Meloe is apparently the parasite of diverse Bees, a suspicion which will be confirmed when we examine the manner in which the larvæ reach the cell full of honey. The Sitares, though less given to change of lodging, are likewise able to inhabit nests of different species. They are very common in the cells of *Anthophora pilipes*; but I have found them also, in very small numbers, it is true, in the cells of *A. personata*.

Despite the presence of *Meloe cicatricosus* in the dwellings of the Mason-bee, which I so often ransacked in compiling the history of the Sitares, I never saw this insect, at any season of the year, wandering on the perpendicular soil, at the entrance of the corridors, for the purpose of laying its eggs there, as the Sitares do; and I should know nothing of the details of the egg-laying if Godart,[2] de Geer[3] and, above all, Newport had not informed us that the Oil-beetles lay their eggs in the earth. According to the last-named author, the various Oil-beetles whom he had the opportunity of observing dig, among the roots of a clump of grass, in a dry soil exposed to the sun, a hole a couple of inches deep which they carefully fill up after laying their eggs there in a heap. This laying is repeated three or four times over, at intervals of a few days during the same season. For each batch of eggs the female digs a special hole, which she does not fail to fill up afterwards. This takes place in April and May.

[2] Jean Baptiste Godart (1775-1823), the principal editor of *L'Histoire naturelle des lépidoptères de France.*—*Translator's Note.*

[3] Baron Karl de Geer (1720-1778), the Swedish entomologist, author of *Mémoires pour servir à l'histoire des insectes* (1752-1778).—*Translator's Note*.

The number of eggs laid in a single batch is really prodigious. In the first batch, which, it is true, is the most prolific of all, *Meloe proscarabæus*, according to Newport's calculations, produces the astonishing number of 4,218 eggs, which is double the number of eggs laid by a Sitaris. And what must the number be, when we allow for the two or three batches that follow the first! The Sitares, entrusting their eggs to the very corridors through which the Anthophora is bound to pass, spare their larvæ a host of dangers which the larvæ of the Meloe have to run, for these, born far from the dwellings of the Bees, are obliged to make their own way to their hymenopterous foster-parents. The Oil-beetles, therefore, lacking the instinct of the Sitares, are endowed with incomparably greater fecundity. The richness of their ovaries atones for the insufficiency of instinct by proportioning the number of germs in accordance with the risks of destruction. What transcendent harmony is this, which thus holds the scales between the fecundity of the ovaries and the perfection of instinct!

The hatching of the eggs takes place at the end of May or in June, about a month after they are laid. The eggs of the Sitares also are hatched after the same lapse of time. But the Meloe-larvæ, more greatly favoured, are able to set off immediately in search of the Bees that are to feed them; while those of the Sitares, hatched in September, have to wait motionless and in complete abstinence for the emergence of the Anthophoræ the entrance to whose cells they guard. I will not describe the young Meloe-larva, which is sufficiently well known, in particular by the description and the diagram furnished by Newport. To enable the reader to understand what follows, I will confine myself to stating that this primary larva is a sort of little yellow louse, long and slender, found in the spring in the down of different Bees.

How has this tiny creature made its way from the underground lodging where the eggs are hatched to the fleece of a Bee? Newport suspects that the young Oil-beetles, on emerging from their natal burrow, climb upon the neighbouring plants, especially upon the Cichoriceæ, and wait, concealed among the petals, until a few Bees chance to plunder the flower, when they promptly fasten on to their fur and allow themselves to be borne away by them. I have more than Newport's suspicions upon this curious point; my personal observations and experiments are absolutely convincing. I will relate them as the first phase of the history of the Bee-louse. They date back to the 23rd of May, 1858.

A vertical bank on the road from Carpentras to Bédoin is this time the scene of my observations. This bank, baked by the sun, is exploited by numerous swarms of Anthophoræ, who, more industrious than their congeners, are in the habit of building, at the entrance to their corridors, with serpentine fillets of earth, a vestibule, a defensive bastion in the form of an arched cylinder. In a word, they are swarms of *A. parietina*. A sparse carpet of turf extends from the edge of the road to the foot of the bank. The more comfortably to follow the work of the Bees, in the hope of wresting some secret from them, I had been lying for a few moments upon this turf, in the very heart of the inoffensive swarm, when my clothes were invaded by legions of little yellow lice, running with desperate eagerness through the hairy thickets of the nap of the cloth. In these tiny creatures, with which I was powdered here and there as with yellow dust, I soon recognized an old acquaintance, the young Oil-beetles, whom I now saw for the first time elsewhere than in the Bees' fur or the interior of their cells. I could not lose so excellent an opportunity of learning how these larvæ manage to establish themselves upon the bodies of their foster-parents.

In the grass where, after lying down for a moment, I had caught these lice were a few plants in blossom, of which the most abundant were three composites: *Hedypnois polymorpha*, *Senecio gallicus* and *Anthemis arvensis*. Now it was on a composite, a dandelion, that Newport seemed to remember seeing some young Oil-beetles; and my attention therefore was first of all directed to the plants which I have named. To my great satisfaction, nearly all the flowers of these three plants, especially those of the camomile (*Anthemis*) were occupied by young Oil-beetles in greater or lesser numbers. On one head of camomile I counted forty of these tiny insects, cowering motionless in the centre of the florets. On the other hand, I could not discover any on the flowers of the poppy or of a wild rocket (*Diplotaxis muralis*) which grew promiscuously among the plants aforesaid. It seems to me, therefore, that it is only on the composite flowers that the Meloe-larvæ await the Bees' arrival.

In addition to this population encamped upon the heads of the composites and remaining motionless, as though it had achieved its object for the moment, I soon discovered yet another, far more numerous, whose anxious activity betrayed a fruitless search. On the ground, in the grass, numberless little larvæ were running in a great flutter, recalling in some respects the tumultuous disorder of an overturned Ant-hill; others were hurriedly climbing to the tip of a blade of grass and descending with the same haste; others again were plunging into the downy fluff of the withered everlastings, remaining there a moment and quickly reappearing to continue their search. Lastly, with a little attention, I was able to convince myself that within an area of a dozen square yards there was perhaps not a single blade of grass which was not explored by several of these larvæ.

I was evidently witnessing the recent emergence of the young Oil-beetles from their maternal lairs. Part of them had already settled on the groundsel- and camomile-flowers to await the arrival of the Bees; but the majority were still wandering in search of this provisional refuge. It was by this wandering population that I had been invaded when I lay down at the foot of the bank. It was impossible that all these larvæ, the tale of whose alarming thousands I would not venture to define, should form one family and recognize a common mother; despite what Newport has told us of the Oil-beetles' astonishing fecundity, I could not believe this, so great was their multitude.

Though the green carpet was continued for a considerable distance along the side of the road, I could not detect a single Meloe-larva elsewhere than in the few square yards lying in front of the bank inhabited by the Mason-bee. These larvæ therefore could not have come far; to find themselves near the Anthophoræ they had had no long pilgrimage to make, for there was not a sign of the inevitable stragglers and laggards that follow in the wake of a travelling caravan. The burrows in which the eggs were hatched were therefore in that turf opposite the Bees' abode. Thus the Oil-beetles, far from laying their eggs at random, as their wandering life might lead one to suppose, and leaving their young to the task of approaching their future home, are able to recognize the spots haunted by the Anthophoræ and lay their eggs in the near neighbourhood of those spots.

With such a multitude of parasites occupying the composite flowers in close proximity to the Anthophora's nests, it is impossible that the majority of the swarm should not become infested sooner or later. At the time of my observations, a comparatively tiny proportion of the starving legion was waiting on the flowers; the others were still wandering on the ground, where the Anthophoræ very rarely alight; and yet I detected the presence of several Meloe-larvæ in the thoracic down of nearly all the Anthophoræ which I caught and examined.

I have also found them on the bodies of the Melecta- and Coelioxys-bees,[4] who are parasitic on the Anthophoræ. Suspending their audacious patrolling before the galleries under construction, these spoilers of the victualled cells alight for an instant on a camomile-flower and lo, the thief is robbed! A tiny, imperceptible louse has slipped into the thick of the downy fur and, at the moment when the parasite, after destroying the Anthophora's egg, is laying her own upon the stolen honey, will creep upon this egg, destroy it in its turn and remain sole mistress of the provisions. The mess of honey amassed by the Anthophora will thus pass through the hands of three owners and remain finally the property of the weakest of the three.

[4] Cf. *The Mason-bees:* chaps. viii. and ix. — *Translator's Note.*

And who shall say whether the Meloe, in its turn, will not be dispossessed by a fresh thief; or even whether it will not, in the state of a drowsy, fat and flabby larva, fall a prey to some marauder who will munch its live entrails? As we meditate upon this deadly, implacable struggle which nature imposes, for their preservation, on these different creatures, which are by turns possessors and dispossessed, devourers and devoured, a painful impression mingles with the wonder aroused by the means employed by each parasite to attain its end; and, forgetting for a moment the tiny world in which these things happen, we are seized with terror at this concatenation of larceny, cunning and brigandage which forms part, alas, of the designs of *alma parens rerum!*

The young Meloe-larvæ established in the down of the Anthophoræ or in that of the Melecta- and the Coelioxys-bees, their parasites, had adopted an infallible means of sooner or later reaching the desired cell. Was it, so far as they were concerned, a choice dictated by the foresight of instinct, or just simply the result of a lucky chance? The question was soon decided. Various Flies—Drone-flies and Bluebottles (*Eristalis tenax* and *Calliphora vomitoria*)—would settle from time to time on the groundsel- or camomile-flowers occupied by the young Meloes and stop for a moment to suck the sweet secretions. On all these Flies, with very few exceptions, I found Meloe-larvæ, motionless in the silky down of the thorax. I may also mention, as infested by these larvæ, an Ammophila (*A. hirsuta*),[5] who victuals her burrows with a caterpillar in early spring, while her kinswomen build their nests in autumn. This Wasp merely grazes, so to speak, the surface of a flower; I catch her; there are Meloes moving about her body. It is clear that neither the Drone-flies nor the Bluebottles, whose larvæ live in putrefying matter, nor yet the Ammophilæ who victual theirs with caterpillars, could ever have carried the larvæ which invaded them into cells filled with honey. These larvæ therefore had gone astray; and instinct, as does not often happen, was here at fault.

[5] For the Wasp known as the Hairy Ammophila, who feeds her young on the Grey Worm, the caterpillar of the Turnip Moth, cf. *The Hunting Wasps*, chaps. xviii. to xx.—*Translator's Note.*

The Glow-Worm and Other Beetles

Let us now turn our attention to the young Meloes waiting expectant upon the camomile-flowers. There they are, ten, fifteen or more, lodged half-way down the florets of a single blossom or in their interstices; it therefore needs a certain degree of scrutiny to perceive them, their hiding-place being the more effectual in that the amber colour of their bodies merges in the yellow hue of the florets. So long as nothing unusual happens upon the flower, so long as no sudden shock announces the arrival of a strange visitor, the Meloes remain absolutely motionless and give no sign of life. To see them dipping vertically, head downwards, into the florets, one might suppose that they were seeking some sweet liquid, their food; but in that case they ought to pass more frequently from one floret to another, which they do not, except when, after a false alarm, they regain their hiding-places and choose the spot which seems to them the most favourable. This immobility means that the florets of the camomile serve them only as a place of ambush, even as later the Anthophora's body will serve them solely as a vehicle to convey them to the Bee's cell. They take no nourishment, either on the flowers or on the Bees; and, as with the Sitares, their first meal will consist of the Anthophora's egg, which the hooks of their mandibles are intended to rip open.

Their immobility is, as we have said, complete; but nothing is easier than to arouse their suspended activity. Shake a camomile-blossom lightly with a bit of straw: instantly the Meloes leave their hiding-places, come up and scatter in all directions on the white petals of the circumference, running over them from one end to the other with all the speed which the smallness of their size permits. On reaching the extreme end of the petals, they fasten to it either with their caudal appendages, or perhaps with a sticky substance similar to that furnished by the anal button of the Sitares; and, with their bodies hanging outside and their six legs free, they bend about in every direction and stretch as far out as they can, as though striving to touch an object out of their reach. If nothing offers for them to seize upon, after a few vain attempts they regain the centre of the flower and soon resume their immobility.

But, if we place near them any object whatever, they do not fail to catch on to it with surprising agility. A blade of grass, a bit of straw, the handle of my tweezers which I hold out to them: they accept anything in their eagerness to quit the provisional shelter of the flower. It is true that, after finding themselves on these inanimate objects, they soon recognize that they have gone astray, as we see by their bustling movements to and fro and their tendency to go back to the flower if there still be time. Those which have thus giddily flung themselves upon a bit of straw and are allowed to return to their flower do not readily fall a second time into the same trap. There is therefore, in these animated specks, a memory, an experience of things.

After these experiments I tried others with hairy materials imitating more or less closely the down of the Bees, with little pieces of cloth or velvet cut from my clothes, with plugs of cotton wool, with pellets of flock gathered from the everlastings. Upon all these objects, offered with the tweezers, the Meloes flung themselves without any difficulty; but, instead of keeping quiet, as they do on the bodies of the Bees, they soon convinced me, by their restless behaviour, that they found themselves as much out of their element on these furry materials as on the smooth surface of a bit of straw. I ought to have expected this: had I not just seen them wandering without pause upon the everlastings enveloped with cottony flock? If reaching the shelter of a downy surface were enough to make them believe themselves safe in harbour, nearly all would perish, without further attempts, in the down of the plants.

Let us now offer them live insects and, first of all, Anthophoræ. If the Bee, after we have rid her of the parasites which she may be carrying, be taken by the wings and held for a moment in contact with the flower, we invariably find her, after this rapid contact, overrun by Meloes clinging to her hairs. The larvæ nimbly take up their position on the thorax, usually on the shoulders or sides, and once there they remain motionless: the second stage of their strange journey is compassed.

The Glow-Worm and Other Beetles

After the Anthophoræ, I tried the first live insects that I was able to procure at once: Drone-flies, Bluebottles, Hive-bees, small Butterflies. All were alike overrun by the Meloes, without hesitation. What is more, there was no attempt made to return to the flowers. As I could not find any Beetles at the moment, I was unable to experiment with them. Newport, experimenting, it is true, under conditions very different from mine, since his observations related to young Meloes held captive in a glass jar, while mine were made in the normal circumstances, Newport, I was saying, saw Meloes fasten to the body of a Malachius and stay there without moving, which inclines me to believe that with Beetles I should have obtained the same results as, for instance, with a Drone-fly. And I did, in fact, at a later date, find some Meloe-larvæ on the body of a big Beetle, the Golden Rose-chafer (*Cetonia aurata*), an assiduous visitor of the flowers.

After exhausting the insect class, I put within their reach my last resource, a large black Spider. Without hesitation they passed from the flower to the arachnid, made for places near the joints of the legs and settled there without moving. Everything therefore seems to suit their plans for leaving the provisional abode where they are waiting; without distinction of species, genus, or class, they fasten to the first living creature that chance brings within their reach. We now understand how it is that these young larvæ have been observed upon a host of different insects and especially upon the early Flies and Bees pillaging the flowers; we can also understand the need for that prodigious number of eggs laid by a single Oil-beetle, since the vast majority of the larvæ which come out of them will infallibly go astray and will not succeed in reaching the cells of the Anthophoræ. Instinct is at fault here; and fecundity makes up for it.

But instinct recovers its infallibility in another case. The Meloes, as we have seen, pass without difficulty from the flower to the objects within their reach, whatever these may be, smooth or hairy, living or inanimate. This done, they behave very differently, according as they have chanced to invade the body of an insect or some other object. In the first case, on a downy Fly or Butterfly, on a smooth-skinned Spider or Beetle, the larvæ remain motionless after reaching the point which suits them. Their instinctive desire is therefore satisfied.

In the second case, in the midst of the nap of cloth or velvet, or the filaments of cotton, or the flock of the everlasting, or, lastly, on the smooth surface of a leaf or a straw, they betray the knowledge of their mistake by their continual coming and going, by their efforts to return to the flower imprudently abandoned.

How then do they recognize the nature of the object to which they have just moved? How is it that this object, whatever the quality of its surface, will sometimes suit them and sometimes not? Do they judge their new lodging by sight? But then no mistake would be possible; the sense of sight would tell them at the outset whether the object within reach was suitable or not; and emigration would or would not take place according to its decision. And then how can we suppose that, buried in the dense thicket of a pellet of cotton-wool or in the fleece of an Anthophora, the imperceptible larva can recognize, by sight, the enormous mass which it is perambulating?

Is it by touch, by some sensation due to the inner vibrations of living flesh? Not so, for the Meloes remain motionless on insect corpses that have dried up completely, on dead Anthophoræ taken from cells at least a year old. I have seen them keep absolutely quiet on fragments of an Anthophora on a thorax long since nibbled and emptied by the Mites. By what sense then can they distinguish the thorax of an Anthophora from a velvety pellet, when sight and touch are out of the question? The sense of smell remains. But in that case what exquisite subtlety must we not take for granted? Moreover, what similarity of smell can we admit between all the insects which, dead or alive, whole or in pieces, fresh or dried, suit the Meloes, while anything else does not suit them? A wretched louse, a living speck, leaves us mightily perplexed as to the sensibility which directs it. Here is yet one more riddle added to all the others.

After the observations which I have described, it remained for me to search the earthen surface inhabited by the Anthophoræ: I should then have followed the Meloe-larva in its transformations. It was certainly *cicatricosus* whose larvæ I had been studying; it was certainly this insect which ravaged the cells of the Mason-bee, for I found it dead in the old galleries which it had been unable to leave.

This opportunity, which did not occur again, promised me an ample harvest. I had to give it all up. My Thursday was drawing to a close; I had to return to Avignon, to resume my lessons on the electrophorus and the Toricellian tube. O happy Thursdays! What glorious opportunities I lost because you were too short!

We will go back a year to continue this history. I collected, under far less favourable conditions, it is true, enough notes to map out the biography of the tiny creature which we have just seen migrating from the camomile-flowers to the Anthophora's back. From what I have said of the Sitaris-larvæ, it is plain that the Meloe-larvæ perched, like the former, on the back of a Bee, have but one aim: to get themselves conveyed by this Bee to the victualled cells. Their object is not to live for a time on the body that carries them.

Were it necessary to prove this, it would be enough to say that we never see these larvæ attempt to pierce the skin of the Bee, or else to nibble at a hair or two, nor do we see them increase in size so long as they are on the Bee's body. To the Meloes, as to the Sitares, the Anthophora serves merely as a vehicle which conveys them to their goal, the victualled cell.

It remains for us to learn how the Meloe leaves the down of the Bee which has carried it, in order to enter the cell. With larvæ collected from the bodies of different Bees, before I was fully acquainted with the tactics of the Sitares, I undertook, as Newport had done before me, certain investigations intended to throw light on this leading point in the Oil-beetle's history. My attempts, based upon those which I had made with the Sitares, resulted in the same failure. The tiny creatures, when brought into contact with Anthophora-larvæ or -nymphs, paid no attention whatever to their prey; others, placed near cells which were open and full of honey, did not enter them, or at most ventured to the edge of the orifice; others, lastly, put inside the cell, on the dry wall or on the surface of the honey, came out again immediately or else got stuck and died. The touch of the honey is as fatal to them as to the young Sitares.

The Glow-Worm and Other Beetles

Searches made at various periods in the nests of the Hairy-footed Anthophora had taught me some years earlier that *Meloe cicatricosus*, like the Sitares, is a parasite of that Bee; indeed I had at different times discovered adult Meloes, dead and shrivelled, in the Bee's cells. On the other hand, I knew from Léon Dufour that the little yellow animal, the Louse found in the Bee's down, had been recognized, thanks to Newport's investigations, as the larva of the Oil-beetle. With these data, rendered still more striking by what I was learning daily on the subject of the Sitares, I went to Carpentras, on the 21st of May, to inspect the nests of the Anthophoræ, then building, as I have described. Though I was almost certain of succeeding, sooner or later, with the Sitares, who were excessively abundant, I had very little hope of the Meloes, which on the contrary are very scarce in the same nests. Circumstances, however, favoured me more than I dared hope and, after six hours' labour, in which the pick played a great part, I became the possessor, by the sweat of my brow, of a considerable number of cells occupied by Sitares and two other cells appropriated by Meloes.

While my enthusiasm had not had time to cool at the sight, momentarily repeated, of a young Sitaris perched upon an Anthophora's egg floating in the centre of the little pool of honey, it might well have burst all restraints on beholding the contents of one of these cells. On the black, liquid honey a wrinkled pellicle is floating; and on this pellicle, motionless, is a yellow louse. The pellicle is the empty envelope of the Anthophora's egg; the louse is a Meloe-larva.

The story of this larva becomes self-evident. The young Meloe leaves the down of the Bee at the moment when the egg is laid; and, since contact with the honey would be fatal to the grub, it must, in order to save itself, adopt the tactics followed by the Sitaris, that is to say, it must allow itself to drop on the surface of the honey with the egg which is in the act of being laid. There, its first task is to devour the egg which serves it for a raft, as is attested by the empty envelope on which it still remains; and it is after this meal, the only one that it takes so long as it retains its present form, that it must commence its long series of transformations and feed upon the honey amassed by the Anthophora. This was the reason of the complete failure both of my attempts and of Newport's to rear the young Meloe-larvæ. Instead of offering them honey, or larvæ, or nymphs, we should have placed them on the eggs recently laid by the Anthophora.

On my return from Carpentras, I meant to try this method, together with that of the Sitares, with which I had been so successful; but, as I had no Meloe-larvæ at my disposal and could not obtain any save by searching for them in the Bees' fleece, the Anthophora-eggs were all discovered to have hatched in the cells which I brought back from my expedition, when I was at last able to find some. This lost experiment is little to be regretted, for, since the Meloes and the Sitares exhibiting the completest similarity not only in habits but also in their method of evolution, there is no doubt whatever that I should have succeeded. I even believe that this method may be attempted with the cells of various Bees, provided that the eggs and the honey do not differ too greatly from the Anthophora's. I should not, for example, count on being successful with the cells of the three-horned Osmia, who shares the Anthophora's quarters: her egg is short and thick; and her honey is yellow, odourless, solid, almost a powder and very faintly flavoured.

CHAPTER V
HYPERMETAMORPHOSIS

By a Machiavellian stratagem the primary larva of the Oil-beetle or the Sitaris has penetrated the Anthophora's cell; it has settled on the egg, which is its first food and its life-raft in one. What becomes of it once the egg is exhausted?

Let us, to begin with, go back to the larva of the Sitaris. By the end of a week the Anthophora's egg has been drained dry by the parasite and is reduced to the envelope, a shallow skiff which preserves the tiny creature from the deadly contact of the honey. It is on this skiff that the first transformation takes place, whereafter the larva, which is now organized to live in a glutinous environment, drops off the raft into the pool of honey and leaves its empty skin, split along the back, clinging to the pellicle of the egg. At this stage we see floating motionless on the honey a milk-white atom, oval, flat and a twelfth of an inch long. This is the larva of the Sitaris in its new form. With the aid of a lens we can distinguish the fluctuations of the digestive canal, which is gorging itself with honey; and along the circumference of the flat, elliptical back we perceive a double row of breathing-pores which, thanks to their position, cannot be choked by the viscous liquid. Before describing the larva in detail we will wait for it to attain its full development, which cannot take long, for the provisions are rapidly diminishing.

The rapidity however is not to be compared with that with which the gluttonous larvæ of the Anthophora consume their food. Thus, on visiting the dwellings of the Anthophoræ for the last time, on the 25th of June, I found that the Bee's larvæ had all finished their rations and attained their full development, whereas those of the Sitares, still immersed in the honey, were, for the most part, only half the size which they must finally attain. This is yet another reason why the Sitares should destroy an egg which, were it to develop, would produce a voracious larva, capable of starving them in a very short time. When rearing the larvæ myself in test-tubes, I have found that the Sitares take thirty-five to forty days to finish their mess of honey and that the larvæ of the Anthophora spend less than a fortnight over the same meal.

It is in the first half of July that the Sitaris-grubs reach their full dimensions. At this period the cell usurped by the parasite contains nothing beyond a full-fed larva and, in a corner, a heap of reddish droppings. This larva is soft and white, about half an inch in length and a quarter of an inch wide at its broadest part. Seen from above as it floats on the honey, it is elliptical in form, tapering gradually towards the front and more suddenly towards the rear. Its ventral surface is highly convex; its dorsal surface, on the contrary, is almost flat. When the larva is floating on the liquid honey, it is as it were steadied by the excessive development of the ventral surface immersed in the honey, which enables it to acquire an equilibrium that is of the greatest importance to its welfare. In fact, the breathing-holes, arranged without means of protection on either edge of the almost flat back, are level with the viscous liquid and would be choked by that sticky glue at the least false movement, if a suitably ballasted hold did not prevent the larva from heeling over. Never was corpulent abdomen of greater use: thanks to this plumpness of the belly the larva is protected from asphyxia.

Its segments number thirteen, including the head. This head is pale, soft, like the rest of the body, and very small compared with the rest of the creature. The antennæ are excessively short and consist of two cylindrical joints. I have vainly looked for the eyes with a powerful magnifying-glass. In its former state, the larva, subject to strange migrations, obviously needs the sense of sight and is provided with four ocelli. In its present state, of what use would eyes be to it at the bottom of a clay cell, where the most absolute darkness prevails?

The labrum is prominent, is not distinctly divided from the head, is curved in front and edged with pale and very fine bristles. The mandibles are small, reddish toward the tips, blunt and hollowed out spoonwise on the inner side. Below the mandibles is a fleshy part crowned with two very tiny nipples. This is the lower lip with its two palpi. It is flanked right and left by two other parts, likewise fleshy, adhering closely to the lip and bearing at the tip a rudimentary palp consisting of two or three very tiny joints. These two parts are the future jaws. All this apparatus of lips and jaws is completely immobile and in a rudimentary condition which is difficult to describe. They are budding organs, still faint and embryonic. The labrum and the complicated lamina formed by the lip and the jaws leave between them a narrow slit in which the mandibles work.

The legs are merely vestiges, for, though they consist of three tiny cylindrical joints, they are barely a fiftieth of an inch in length. The creature is unable to make use of them, not only in the liquid honey upon which it lives, but even on a solid surface. If we take the larva from the cell and place it on a hard substance, to observe it more readily, we see that the inordinate protuberance of the abdomen, by lifting the thorax from the ground, prevents the legs from finding a support. Lying on its side, the only possible position because of its conformation, the larva remains motionless or only makes a few lazy, wriggling movements of the abdomen, without ever stirring its feeble limbs, which for that matter could not assist it in any way. In short, the tiny creature of the first stage, so active and alert, is succeeded by a ventripotent grub, deprived of movement by its very obesity. Who would recognize in this clumsy, flabby, blind, hideously pot-bellied creature, with nothing but a sort of stumps for legs, the elegant pigmy of but a little while back, armour-clad, slender and provided with highly perfected organs for performing its perilous journeys?

Lastly, we count nine pairs of stigmata: one pair on the mesothorax and the rest on the first eight segments of the abdomen. The last pair, that on the eighth abdominal segment, consists of stigmata so small that to detect them we have to gather their position by that in the succeeding states of the larva and to pass a very patient magnifying-glass along the direction of the other pairs. These are as yet but vestigial stigmata. The others are fairly large, with pale, round, flat edges.

If in its first form the Sitaris-larva is organized for action, to obtain possession of the coveted cell, in its second form it is organized solely to digest the provisions acquired. Let us take a glance at its internal structure and in particular at its digestive apparatus. Here is a strange thing: this apparatus, in which the hoard of honey amassed by the Anthophora is to be engulfed, is similar in every respect to that of the adult Sitaris, who possibly never takes food. We find in both the same very short oesophagus, the same chylific ventricle, empty in the perfect insect, distended in the larva with an abundant orange-coloured pulp; in both the same gall-bladders, four in number, connected with the rectum by one of their extremities. Like the perfect insect, the larva is devoid of salivary glands or any other similar apparatus. Its nervous system comprises eleven ganglia, not counting the oesophageal collar, whereas in the perfect insect there are only seven: three for the thorax, of which the last two are contiguous, and four for the abdomen.

When its rations are finished the larva remains a few days in a motionless condition, ejecting from time to time a few reddish droppings until the digestive canal is completely cleared of its orange-coloured pulp. Then the creature contracts itself, huddles itself together; and before long we see coming detached from its body a transparent, slightly crumpled and extremely fine pellicle, forming a closed bag, in which the successive transformations will take place henceforth. On this epidermal bag, this sort of transparent leather bottle, formed by the larva's skin detached all of a piece, without a slit of any kind, we can distinguish the several well-preserved external organs: the head, with its antennæ, mandibles, paws and palpi; the thoracic segments, with their vestiges of legs; the abdomen, with its chain of breathing-holes still connected one to another by tracheal threads.

Then beneath this pellicle, which is so delicate that it can hardly bear the most cautious touch, we see a soft, white mass taking shape, a mass which in a few hours acquires a firm, horny consistency and a vivid yellow hue. The transformation is now complete. Let us tear the fine gauze bag enclosing the organism which has just come into being and direct our investigation to this third form of the Sitaris-larva.

The Glow-Worm and Other Beetles

It is an inert, segmented body, with an oval outline, a horny consistency, just like that of pupæ and chrysalids, and a bright-yellow colour, which we can best describe by likening it to that of a lemon-drop. Its upper surface forms a double inclined plane with a very blunt ridge; its lower surface is at first flat, but, as the result of evaporation, becomes more concave daily, leaving a projecting rim all around its oval outline. Lastly, its two extremities or poles are slightly flattened. The major axis of the lower surface averages half an inch in length and the minor axis a quarter of an inch.

At the cephalic pole of this body is a sort of mask, modelled roughly on the head of the larva, and at the opposite pole a small circular disk deeply wrinkled at the centre. The three segments that come after the head bear each a pair of very minute knobs, hardly visible without the lens: these are, to the legs of the larva in its previous form, what the cephalic mask is to the head of the same larva. They are not organs, but indications, landmarks placed at the points where these organs will appear later. On either side we count nine stigmata, set as before on the mesothorax and the first eight abdominal segments. The first eight breathing-holes are dark brown and stand out plainly against the yellow colour of the body. They consist of small, shiny, conical knobs, perforated at the top with a round hole. The ninth stigma, though fashioned like the others, is ever so much smaller; it cannot be distinguished without the lens.

The anomaly, already so manifest in the change from the first form to the second, becomes even more so here; and we do not know what name to give to an organism without a standard of comparison, not only in the order of Beetles, but in the whole class of insects. While, on the one hand, this organism offers many points of resemblance to the pupæ of the Flies in its horny consistency, in the complete immobility of its various segments, in the all but absolute absence of relief which would enable one to distinguish the parts of the perfect insect; while, on the other hand, it approximates to the chrysalids, because the creature, to attain this condition, has to shed its skin, as the caterpillars do, it differs from the pupa because it has for covering not the surface skin, which has become horny, but rather one of the inner skins of the larva; and it differs from the chrysalids by the absence of mouldings which in the latter betray the appendages of the perfect insect. Lastly, it differs yet more profoundly from the pupa and the chrysalis because from both these organisms the perfect insect springs straightway, whereas that which follows what we are considering is simply a larva like that which went before. I shall suggest, to denote this curious organism, the term *pseudochrysalis;* and I shall reserve the names *primary larva*, *secondary larva* and *tertiary larva* to denote, in a couple of words, each of the three forms under which the Sitares possess all the characteristics of larvæ.

Although the Sitaris, on assuming the form of the pseudochrysalis, is transfigured outwardly to the point of baffling the science of entomological phases, this is not so inwardly. I have at every season of the year examined the viscera of the pseudochrysalids, which generally remain stationary for a whole year, and I have never observed other forms among their organs than those which we find in the secondary larva. The nervous system has undergone no change. The digestive apparatus is absolutely void and, because of its emptiness, appears only as a thin cord, sunk, lost amid the adipose sacs. The stercoral intestine has more substance; its outlines are better defined. The four gall-bladders are always perfectly distinct. The adipose tissue is more abundant than ever: it forms by itself the whole contents of the pseudochrysalis, for in the matter of volume the insignificant threads of the nervous system and the digestive apparatus count for nothing. It is the reserve upon which life must draw for its future labours.

A few Sitares remain hardly a month in the pseudochrysalis stage. The other phases are achieved in the course of August; and at the beginning of September the insect attains the perfect state. But as a rule the development is slower; the pseudochrysalis goes through the winter; and it is not, at the earliest, until June in the second year that the final transformations take place. Let us pass in silence over this long period of repose, during which the Sitaris, in the form of a pseudochrysalis, slumbers at the bottom of its cell, in a sleep as lethargic as that of a germ in its egg, and come to the months of June and July in the following year, the period of what we might call a second hatching.

The pseudochrysalis is still enclosed in the delicate pouch formed of the skin of the secondary larva. Outside, nothing fresh has happened; but important changes have taken place inside. I have said that the pseudochrysalis displayed an upper surface arched like a hog's back and a lower surface at first flat and then more and more concave. The sides of the double inclined plane of the upper or dorsal surface also share in this depression occasioned by the evaporation of the fluid constituents; and a time comes when these sides are so depressed that a section of the pseudochrysalis through a plane perpendicular to its axis would be represented by a curvilinear triangle with blunted corners and inwardly convex sides. This is the appearance displayed by the pseudochrysalis during the winter and spring.

But in June it has lost this withered appearance; it represents a perfect balloon, an ellipsoid of which the sections perpendicular to the major axis are circles. Something has also come to pass of greater importance than this expansion, which may be compared with that which we obtain by blowing into a wrinkled bladder. The horny integuments of the pseudochrysalis have become detached from their contents, all of a piece, without a break, just as happened the year before with the skin of the secondary larva; and they thus form a fresh vesicular envelope, free from any adhesion to the contents and itself enclosed in the pouch formed of the secondary larva's skin. Of these two bags without outlet, one of which is enclosed within the other, the outer is transparent, flexible, colourless and extremely delicate; the second is brittle, almost as delicate as the first, but much less translucent because of its yellow colouring, which makes it resemble a thin flake of amber. On this second sac are found the stigmatic warts, the thoracic studs and so forth, which we noted on the pseudochrysalis. Lastly, within its cavity we catch a glimpse of something the shape of which at once recalls to mind the secondary larva.

And indeed, if we tear the double envelope which protects this mystery, we recognize, not without astonishment, that we have before our eyes a new larva similar to the secondary. After one of the strangest transformations, the creature has gone back to its second form. To describe the new larva is unnecessary, for it differs from the former in only a few slight details. In both there is the same head, with its various appendages barely outlined; the same vestiges of legs, the same stumps transparent as crystal. The tertiary larva differs from the secondary only by its abdomen, which is less fat, owing to the absolute emptiness of the digestive apparatus; by a double chain of fleshy cushions extending along each side; by the rim of the stigmata, crystalline and slightly projecting, but less so than in the pseudochrysalis; by the ninth pair of breathing-holes, hitherto rudimentary but now almost as large as the rest; lastly by the mandibles ending in a very sharp point. Evicted from its twofold sheath, the tertiary larva makes only very lazy movements of contraction and dilation, without being able to advance, without even being able to maintain its normal position, because of the weakness of its legs. It usually remains motionless, lying on its side, or else displays its drowsy activity merely by feeble, wormlike movements.

By dint of these alternate contractions and dilations, indolent though they be, the larva nevertheless contrives to turn right round in the sort of shell with which the pseudochrysalidal integuments provide it, when by accident it finds itself placed head downwards; and this operation is all the more difficult inasmuch as the larva almost exactly fills the cavity of the shell. The creature contracts, bends its head under its belly and slides its front half over its hinder half by wormlike movements so slow that the lens can hardly detect them. In less than a quarter of an hour the larva, at first turned upside down, finds itself again head uppermost. I admire this gymnastic feat, but have some difficulty in understanding it, so small is the space which the larva, when at rest in its cell, leaves unoccupied, compared with that which we should be justified in expecting from the possibility of such a reversal. The larva does not long enjoy the privilege which enables it to resume inside its cell, when this is moved from its original position, the attitude which it prefers, that is to say, with its head up.

Two days, at most, after its first appearance it relapses into an inertia as complete as that of the pseudochrysalis. On removing it from its amber shell, we see that its faculty of contracting or dilating at will is so completely paralysed that the stimulus of a needle is unable to provoke it, though the integuments have retained all their flexibility and though no perceptible change has occurred in the organization. The irritability, therefore, which in the pseudochrysalis is suspended for a whole year, reawakens for a moment, to relapse instantly into the deepest torpor. This torpor will be partly dispelled only at the moment of the passing into the nymphal stage, to return immediately afterwards and last until the insect attains the perfect state.

Further, on holding larvæ of the third form, or nymphs enclosed in their cells, in an inverted position, in glass tubes, we never see them regain an erect position, however long we continue the experiment. The perfect insect itself, during the time that it is enclosed in the shell, cannot regain it, for lack of the requisite flexibility. This total absence of movement in the tertiary larva, when a few days old, and also in the nymph, together with the smallness of the space left free in the shell, would necessarily lead to the conviction, if we had not witnessed the first moments of the tertiary larva, that it is absolutely impossible for the creature to turn right round.

And now see to what curious inferences this lack of observations made at the due moment may lead us. We collect some pseudochrysalids and heap them in a glass jar in all possible positions. The favourable season arrives; and with very legitimate astonishment we find that, in a large number of shells, the larva or nymph occupies an inverted position, that is to say, the head is turned towards the anal extremity of the shell. In vain we watch these reversed bodies for any indications of movement; in vain we place the shells in every imaginable position, to see if the creature will turn round; in vain, once more, we ask ourselves where the free space is which this turning would demand. The illusion is complete:

I have been taken in by it myself; and for two years I indulged in the wildest conjectures to account for this lack of correspondence between the shell and its contents, to explain, in short, a fact which is inexplicable once the propitious moment has passed.

On the natural site, in the cells of the Anthophora, this apparent anomaly never occurs, because the secondary larva, when on the point of transformation into the pseudochrysalis, is always careful to place its head uppermost, according as the axis of the cell more or less nearly approaches the vertical. But, when the pseudochrysalids are placed higgledy-piggledy in a box or jar, all those which are upside down will later contain inverted larvæ or nymphs.

After four changes of form so profound as those which I have described, one might reasonably expect to find some modifications of the internal organization. Nevertheless, nothing is changed; the nervous system is the same in the tertiary larva as in the earlier phases; the reproductive organs do not yet show; and there is no need to mention the digestive apparatus, which remains invariable even in the perfect insect.

The duration of the tertiary larva is a bare four or five weeks, which is also about the duration of the second. In July, when the secondary larva passes into the pseudochrysalid stage, the tertiary larva passes into the nymphal stage, still inside the double vesicular envelope. Its skin splits along the back in front; and with the assistance of a few feeble contractions, which reappear at this juncture, it is thrust behind in the shape of a little ball. There is therefore nothing here that differs from what happens in the other Beetles.

Nor does the nymph which succeeds this tertiary larva present any peculiarity: it is the perfect insect in swaddling-bands, yellowish white, with its various external members, clear as crystal, displayed under the abdomen. A few weeks elapse, during which the nymph partly dons the livery of the adult state; and, in about a month, the insect moults for a last time, in the usual manner, in order to attain its final form. The wing-cases are now of a uniform yellowish white, as are the wings, the abdomen and the greater part of the legs; very nearly all the rest of the body is of a glossy black. In the space of twenty-four hours, the wing-cases assume their half-black, half-russet colouring; the wings grow darker; and the legs finish turning black. This done, the adult organism is completed. However, the Sitaris remains still a fortnight in the intact shell, ejecting at intervals white droppings of uric acid, which it pushes back together with the shreds of its last two sloughs, those of the tertiary larva and of the nymph. Lastly, about the middle of August, it tears the double bag that contains it, pierces the lid of the Anthophora's cell, enters a corridor and appears outside in quest of the other sex.

I have told how, while digging in search of the Sitaris, I found two cells belonging to *Meloe cicatricosus*. One contained an Anthophora's egg; with this egg was a yellow Louse, the primary larva of the Meloe. The history of this tiny creature we know. The second cell also was full of honey. On the sticky liquid floated a little white larva, about a sixth of an inch in length and very different from the other little white larvæ belonging to Sitares. The rapid fluctuations of the abdomen showed that it was eagerly drinking the strong-scented nectar collected by the Bee. This larva was the young Meloe in the second period of its development.

I was not able to preserve these two precious cells, which I had opened wide to examine the contents. On my return from Carpentras, I found that their honey had been spilt by the motion of the carriage and that their inhabitants were dead. On the 25th of June, a fresh visit to the nests of the Anthophoræ furnished me with two larvæ like the foregoing, but much larger. One of them was on the point of finishing its store of honey, the other still had nearly half left. The first was put in a place of safety with a thousand precautions, the second was at once immersed in alcohol.

These larvæ are blind, soft, fleshy, yellowish white, covered with a fine down visible only under the lens, curved into a fish-hook like the larvæ of the Lamellicorns, to which they bear a certain resemblance in their general configuration. The segments, including the head, number thirteen, of which nine are provided with breathing-holes with a pale, oval rim. These are the mesothorax and the first eight abdominal segments. As in the Sitaris-larvæ, the last pair of stigmata, that of the eighth segment of the abdomen, is less developed than the rest.

The head is horny, of a light brown colour. The epistoma is edged with brown. The labrum is prominent, white and trapezoidal. The mandibles are black, strong, short, obtuse, only slightly curved, sharp-edged and furnished each with a broad tooth on the inner side. The maxillary and labial palpi are brown and shaped like very small studs with two or three joints to them. The antennæ, inserted just at the base of the mandibles, are brown, and consist of three sections: the first is thick and globular; the two others are much smaller in diameter and cylindrical. The legs are short, but fairly strong, able to serve the creature for crawling or digging; they end in a strong black claw. The length of the larva when fully developed is one inch.

As far as I can judge from the dissection of the specimen preserved in alcohol, whose viscera were affected by being kept too long in that liquid, the nervous system consists of eleven ganglia, not counting the oesophageal collar; and the digestive apparatus does not differ perceptibly from that of an adult Oil-beetle.

The larger of the two larvæ of the 25th of June, placed in a test-tube with what remained of its provisions, assumed a new form during the first week of the following month. Its skin split along the front dorsal half and, after being pushed half back, left partly uncovered a pseudochrysalis bearing the closest analogy with that of the Sitares. Newport did not see the larva of the Oil-beetle in its second form, that which it displays when it is eating the mess of honey hoarded by the Bees, but he did see its moulted skin half-covering the pseudochrysalis which I have just mentioned. From the sturdy mandibles and the legs armed with a powerful claw which he observed on this moulted skin, Newport assumed that, instead of remaining in the same Anthophora-cell, the larva, which is capable of burrowing, passes from one cell to another in search of additional nourishment. This suspicion seems to me to be well-founded, for the size which the larva finally attains exceeds the proportions which the small quantity of honey enclosed in a single cell would lead us to expect.

Let us go back to the pseudochrysalis. It is, as in the Sitares, an inert body, of a horny consistency, amber-coloured and divided into thirteen segments, including the head. Its length is 20 millimetres.[1] It is slightly curved into an arc, highly convex on the dorsal surface, almost flat on the ventral surface and edged with a projecting fillet which marks the division between the two. The head is only a sort of mask on which certain features are vaguely carved in still relief, corresponding with the future parts of the head. On the thoracic segments are three pairs of tubercles, corresponding with the legs of the recent larva and the future insect. Lastly, there are nine pairs of stigmata, one pair on the mesothorax and the eight following pairs on the first eight segments of the abdomen. The last pair is rather smaller than the rest, a peculiarity which we have already noted in the larva which precedes the pseudochrysalis.

[1] .787 inch. — *Translator's Note.*

On comparing the pseudochrysalids of the Oil-beetles and Sitares, we observe a most striking similarity between the two. The same structure occurs in both, down to the smallest details. We find on either side the same cephalic masks, the same tubercles occupying the place of the legs, the same distribution and the same number of stigmata and, lastly, the same colour, the same rigidity of the integuments. The only points of difference are in the general appearance, which is not the same in the two pseudochrysalids, and in the covering formed by the cast skin of the late larva. In the Sitares, in fact, this cast skin constitutes a closed bag, a pouch completely enveloping the pseudochrysalis; in the Oil-beetles, on the contrary, it is split down the back and pushed to the rear and, consequently, only half-covers the pseudochrysalis.

The post-mortem examination of the only pseudochrysalis in my possession showed me that, similarly to that which happens in the Sitares, no change occurred in the organization of the viscera, notwithstanding the profound transformations which take place externally. In the midst of innumerable little sacs of adipose tissue is buried a thin thread in which we easily recognize the essential features of the digestive apparatus, both of the preceding larval form and of the perfect insect. As for the medullary cord of the abdomen, it consists, as in the larva, of eight ganglia. In the perfect insect it comprises only four.

I could not say positively how long the Oil-beetle remains in the pseudochrysalid form; but, if we consider the very complete analogy between the evolution of the Oil-beetles and that of the Sitares, there is reason to believe that a few pseudochrysalids complete their transformation in the same year, while others, in greater numbers, remain stationary for a whole year and do not attain the state of the perfect insect until the following spring. This is also the opinion expressed by Newport.

Be this as it may, I found at the end of August one of these pseudochrysalids which had already attained the nymphal stage. It is with the help of this precious capture that I shall be able to finish the story of the Oil-beetle's development. The horny integuments of the pseudochrysalis are split along a fissure which includes the whole ventral surface and the whole of the head and runs up the back of the thorax. This cast skin, which is stiff and keeps its shape, is half-enclosed, as was the pseudochrysalis, in the skin shed by the secondary larva. Lastly, through the fissure, which divides it almost in two, a Meloe-nymph half-emerges; so that, to all appearances, the pseudochrysalis has been followed immediately by the nymph, which does not happen with the Sitares, which pass from the first of these two states to the second only by assuming an intermediary form closely resembling that of the larva which eats the store of honey.

But these appearances are deceptive, for, on removing the nymph from the split sheath formed by the integuments of the pseudochrysalis, we find, at the bottom of this sheath, a third cast skin, the last of those which the creature has so far rejected. This skin is even now adhering to the nymph by a few tracheal filaments. If we soften it in water, we easily recognize that it possesses an organization almost identical with that which preceded the pseudochrysalis. In the latter case only, the mandibles and the legs are not so robust. Thus, after passing through the pseudochrysalid stage, the Oil-beetles for some time resume the preceding form, almost without modification.

The nymph comes next. It presents no peculiarities. The only nymph that I have reared attained the perfect insect state at the end of September. Under ordinary conditions would the adult Oil-beetle have emerged from her cell at this period? I do not think so, since the pairing and egg-laying do not take place until the beginning of spring. She would no doubt have spent the autumn and the winter in the Anthophora's dwelling, only leaving it in the spring following. It is even probable that, as a rule, the development is even slower and that the Oil-beetles, like the Sitares, for the most part spend the cold season in the pseudochrysalid state, a state well-adapted to the winter torpor, and do not achieve their numerous forms until the return of the warm weather.

The Sitares and Meloes belong to the same family, that of the Meloidæ.[2] Their strange transformations must probably extend throughout the group; indeed, I had the good fortune to discover a third example, which I have not hitherto been able to study in all its details after twenty-five years of investigation. On six occasions, no oftener, during this long period I have set eyes on the pseudochrysalis which I am about to describe. Thrice I obtained it from old Chalicodoma-nests built upon a stone, nests which I at first attributed to the Chalicodoma of the Walls and which I now refer with greater probability to the Chalicodoma of the Sheds. I once extracted it from the galleries bored by some wood-eating larva in the trunk of a dead wild pear-tree, galleries afterwards utilized for the cells of an Osmia, I do not know which. Lastly, I found a pair of them in between the row of cocoons of the Three-pronged Osmia (*O. tridentata*, DUF.), who provides a home for her larvæ in a channel dug in the dry bramble stems. The insect in question therefore is a parasite of the Osmiæ. When I extract it from the old Chalicodoma-nests, I have to attribute it not to this Bee but to one of the Osmiæ (*O.*

tricornis and *O. Latreillii*) who, when making their nests, utilize the old galleries of the Mason-bee.

[2] Later classifiers place both in the family of the Cantharidæ.—*Translator's Note.*

The most nearly complete instances that I have seen furnishes me with the following data: the pseudochrysalis is very closely enveloped in the skin of the secondary larva, a skin consisting of fine transparent pellicle, without any rent whatever. This is the pouch of the Sitaris, save that it lies in immediate contact with the body enclosed. On this jacket we distinguish three pairs of tiny legs, reduced to short vestiges, to stumps. The head is in place, showing quite perceptibly the fine mandibles and the other parts of the mouth. There is no trace of eyes. Each side has a white edging of shrivelled tracheæ, running from one stigmatic orifice to another.

Next comes the pseudochrysalis, horny, currant-red, cylindrical, cone-shaped at both ends, slightly convex on the dorsal surface and concave on the ventral surface. It is covered with delicate, prominent spots, sprinkled very close together; it takes a lens to show them. It is 1 centimetre long and 4 millimetres wide.[3] We can distinguish a large knob of a head, on which the mouth is vaguely outlined; three pairs of little shiny brown specks, which are the hardly perceptible vestiges of the legs; and on each side a row of eight black specks, which are the stigmatic orifices. The first speck stands by itself, in front; the seven others, divided from the first by an empty space, form a continuous row. Lastly, at the opposite end is a little pit, the sign of the anal pore.

[3] .393 x .156 inch.—*Translator's Note.*

Of the six pseudochrysalids which a lucky accident placed at my disposal, four were dead; the other two were furnished by *Zonitis mutica*. This justified my forecast, which from the first, with analogy for my guide, made me attribute these curious organizations to the genus Zonitis. The meloidal parasite of the Osmiæ, therefore, is recognized. We have still to make the acquaintance of the primary larva, which gets itself carried by the Osmia into the cell full of honey, and the tertiary larva, the one which, at a given moment, must be found contained in the pseudochrysalis, a larva which will be succeeded by the nymph.

Let us recapitulate the strange metamorphoses which I have sketched. Every Beetle-larva, before attaining the nymphal stage, undergoes a greater or smaller number of moults, of changes of skin; but these moults, which are intended to favour the development of the larva by ridding it of covering that has become too tight for it, in no way alter its external shape. After any moult that it may have undergone, the larva retains the same characteristics. If it begin by being tough, it will not become tender; if it be equipped with legs, it will not be deprived of them later; if it be provided with ocelli, it will not become blind. It is true that the diet of these non-variable larvæ remains the same throughout their duration, as do the conditions under which they are destined to live.

But suppose that this diet varies, that the environment in which they are called upon to live changes, that the circumstances accompanying their development are liable to great changes: it then becomes evident that the moult may and even must adapt the organization of the larva to these new conditions of existence. The primary larva of the Sitaris lives on the body of the Anthophora. Its perilous peregrinations demand agility of movement, long-sighted eyes and masterly balancing-appliances; it has, in fact, a slender shape, ocelli, legs and special organs adapted to averting a fall. Once inside the Bee's cell, it has to destroy the egg; its sharp mandibles, curved into hooks, will fulfil this office. This done, there is a change of diet: after the Anthophora's egg the larva proceeds to consume the ration of honey. The environment in which it has to live also changes: instead of balancing itself on a hair of the Anthophora, it has now to float on a sticky fluid; instead of living in broad daylight, it has to remain plunged in the profoundest darkness. Its sharp mandibles must therefore become hollowed into a spoon that they may scoop up the honey; its legs, its cirri, its balancing-appliances must disappear as useless and even harmful, since all these organs can only involve the larva in serious danger, by causing it to stick in the honey; its slender shape, its horny integuments, its ocelli, being no longer necessary in a dark cell where movement is impossible, where there are no rough encounters to be feared, may likewise give place to complete blindness, to soft integuments, to a heavy, slothful form. This transfiguration, which everything shows to be indispensable to the life of the larva, is effected by a simple moult.

We do not so plainly perceive the necessity of the subsequent forms, which are so abnormal that nothing like them is known in all the rest of the insect class. The larva which is fed on honey first adopts a false chrysalid appearance and afterwards goes back to its earlier form, though the necessity for these transformations escapes us entirely. Here I am obliged to record the facts and to leave the task of interpreting them to the future. The larva of the Meloidæ, therefore, undergo four moults before attaining the nymphal state; and after each moult their characteristics alter most profoundly. During all these external changes, the internal organization remains unchangingly the same; and it is only at the moment of the nymph's appearance that the nervous system becomes concentrated and that the reproductive organs are developed, absolutely as in the other Beetles.

Thus, to the ordinary metamorphoses which make a Beetle pass successively through the stages of larva, nymph and perfect insect, the Meloidæ add others which repeatedly transform the larva's exterior, without introducing any modification of its viscera. This mode of development, which preludes the customary entomological forms by the multiple transfigurations of the larva, certainly deserves a special name: I suggest that of *hypermetamorphosis*.

Let us now recapitulate the more prominent facts of this essay.

The Sitares, the Meloes, the Zonites and apparently other Meloidæ, possibly all of them, are in their earliest infancy parasites of the harvesting Bees.

The larva of the Meloidæ, before reaching the nymphal state, passes through four forms, which I call the *primary larva*, the *secondary larva*, the *pseudochrysalis* and the *tertiary larva*. The passage from one of these forms to the next is effected by a simple moult, without any changes in the viscera.

The primary larva is leathery and settles on the Bee's body. Its object is to get itself carried into a cell filled with honey. On reaching the cell, it devours the Bee's egg; and its part is played.

The secondary larva is soft and differs completely from the primary larva in its external characteristics. It feeds upon the honey contained in the usurped cell.

The pseudochrysalis is a body deprived of all movement and clad in horny integuments which may be compared with those of the pupæ and chrysalids. On these integuments we see a cephalic mask without distinct or movable parts, six tubercles indicating the legs and nine pairs of breathing-holes. In the Sitares the pseudochrysalis is enclosed in a sort of sealed pouch and in the Zonites in a tight-fitting bag formed of the skin of the secondary larva. In the Meloes it is simply half-sheathed in the split skin of the secondary larva.

The tertiary larva reproduces almost exactly the peculiarities of the second; it is enclosed, in the Sitares and probably also the Zonites, in a double vesicular envelope formed of the skin of the secondary larva and the slough of the pseudochrysalis. In the Meloes, it is half-enclosed in the split integuments of the pseudochrysalis, even as these, in their turn, are half-enclosed in the skin of the secondary larva.

From the tertiary larva onwards the metamorphoses follow their habitual course, that is to say, this larva becomes a nymph; and this nymph the perfect insect.

CHAPTER VI
CEROCOMÆ, MYLABRES AND ZONITES

All has not been told concerning the Meloidæ, those strange parasites, some of which, the Sitares and the Oil-beetles, attach themselves, like the tiniest of Lice, to the fleece of various Bees to get themselves carried into the cell where they will destroy the egg and afterwards feed upon the ration of honey. A most unexpected discovery, made a few hundred yards from my door, has warned me once again how dangerous it is to generalize. To take it for granted, as the mass of data hitherto collected seemed to justify us in doing, that all the Meloidæ of our country usurp the stores of honey accumulated by the Bees, was surely a most judicious and natural generalization. Many have accepted it without hesitation; and I for my part was one of them. For on what are we to base our conviction when we imagine that we are stating a law? We think to take our stand upon the general; and we plunge into the quicksands of error. And behold, the law of the Meloidæ has to be struck off the statutes, a fate common to many others, as this chapter will prove.

On the 16th of July, 1883, I was digging, with my son Émile, in the sandy heap where, a few days earlier, I had been observing the labours and the surgery of the Mantis-killing Tachytes. My purpose was to collect a few cocoons of this Digger-wasp. The cocoons were turning up in abundance under my pocket-trowel, when Émile presented me with an unknown object. Absorbed in my task of collection, I slipped the find into my box without examining it further than with a rapid glance. We left the spot. Half-way home, the ardour of my search became assuaged; and a thought of the problematical object, so negligently dropped into the box among the cocoons, flashed across my mind.

"Hullo!" I said to myself. "Suppose it were *that*? Why not? But, no, yes, it *is* that; that's just what it is!"

Then, suddenly turning to Émile, who was rather surprised by this soliloquy:

"My boy," I said, "you have had a magnificent find. It's a pseudochrysalis of the Meloidæ. It's a document of incalculable value; you've struck a fresh vein in the extraordinary records of these creatures. Let us look at it closely and at once."

The thing was taken from the box, dusted by blowing on it and carefully examined. I really had before my eyes the pseudochrysalis of some Meloid. Its shape was unfamiliar to me. No matter: I was an old hand and could not mistake its source. Everything assured me that I was on the track of an insect that rivalled the Sitares and the Oil-beetles in the strangeness of its transformations; and, what was a still more precious fact, its occurrence amid the burrows of the Mantis-killer told me that its habits would be wholly different.

"It's very hot, my poor Émile; we are both of us pretty done. Never mind: let's go back to our sand-hill and dig and have another search. I must have the larva that comes before the pseudochrysalis; I must, if possible, have the insect that comes out of it."

Success responded amply to our zeal. We found a goodly number of pseudochrysalids. More often still, we unearthed larvæ which were busy eating the Mantes, the rations of the Tachytes. Are these really the larvæ that turn into the pseudochrysalids? It seems very probable, but there is room for doubt. Rearing them at home will dispel the mists of probability and replace them by the light of certainty. But that is all: I have not a vestige of the perfect insect to inform me of the nature of the parasite. The future, let us hope, will fill this gap. Such was the result of the first trench opened in the heap of sand. Later searches enriched my harvest a little, without furnishing me with fresh data.

Let us now proceed to examine my double find. And first of all the pseudochrysalis, which put me on the alert. It is a motionless, rigid body, of a waxen yellow, smooth, shiny, curved like a fish-hook towards the head, which is inflected. Under a very powerful magnifying-glass the surface is seen to be strewn with very tiny points which are slightly raised and shinier than the surface. There are thirteen segments, including the head. The dorsal surface is convex, the ventral surface flat. A blunt ridge divides the two surfaces. The three thoracic segments bear each a pair of tiny conical nipples, of a deep rusty red, signs of the future legs. The stigmata are very distinct, appearing as specks of a deeper red than the rest of the integuments. There is one pair, the largest, on the second segment of the thorax, almost on the line dividing it from the first segment. Then follow eight pairs, one on each segment of the abdomen except the last, making in all nine pairs of stigmata. The last pair, that of the eighth abdominal segment, is the smallest.

The anal extremity displays no peculiarity. The cephalic mask comprises eight cone-shaped tubercles, dark red like the tubercles of the legs. Six of these are arranged in two lateral rows; the others are between the two rows. In each row of three nipples, the one in the middle is the largest; it no doubt corresponds with the mandibles. The length of this organism varies greatly, fluctuating between 8 and 15 millimetres.[1] Its width is from 3 to 4 millimetres.[2]

[1] .312 to .585 inch. — *Translator's Note.*

[2] .117 to .156 inch. — *Translator's Note.*

Apart from the general configuration, it will be seen that we have here the strikingly characteristic appearance of the pseudochrysalids of the Sitares, Oil-beetles and Zonites. There are the same rigid integuments, of the red of a cough-lozenge or virgin wax; the same cephalic mask, in which the future mouth-parts are represented by faintly marked tubercles; the same thoracic studs, which are the vestiges of the legs; the same distribution of the stigmata. I was therefore firmly convinced that the parasite of the Mantis-hunters could only be a Meloid.

Let us also record the description of the strange larva found devouring the heap of Mantes in the burrows of the Tachytes. It is naked, blind, white, soft and sharply curved. Its general appearance suggests the larva of some Weevil. I should be even more accurate if I compared it with the secondary larva of *Meloe cicatricosus*, of which I once published a drawing in the *Annales des sciences naturelles*.[3] If we reduce the dimensions considerably, we shall have something very like the parasite of the Tachytes.

[3] It was his essays in this periodical, on the metamorphoses of the Sitares and Oil-beetles, that procured Fabre his first reputation as an entomologist.—*Translator's Note*.

The head is large, faintly tinged with red. The mandibles are strong, bent into a pointed hook, black at the tip and a fiery red at the base. The antennæ are very short, inserted close to the root of the mandibles. I count three joints: the first thick and globular, the other two cylindrical, the second of these cut short abruptly. There are twelve segments, apart from the head, divided by fairly definite grooves. The first thoracic segment is a little longer than the rest, with the dorsal plate very slightly tinged with russet, as is the top of the head. Beginning with the tenth segment, the body tapers a little. A slight scalloped rim divides the dorsal from the ventral surface.

The legs are short, white and transparent and end in a feeble claw. A pair of stigmata on the mesothorax, near the line of junction with the prothorax; a stigma on either side of the first eight abdominal segments; in all nine pairs of stigmata, distributed like those of the pseudochrysalis. These stigmata are small, tinged with red and rather difficult to distinguish. Varying in size, like the pseudochrysalid which seems to come from it, this larva averages nearly half an inch in length and an eighth of an inch in width.

The six little legs, feeble though they be, perform services which one would not at first suspect. They embrace the Mantis that is being devoured and hold her under the mandibles, while the grub, lying on its side, takes its meal at its ease. They also serve for locomotion. On a firm surface, such as the wooden top of my table, the larva can move about quite well; it toddles along, dragging its belly, with its body straight from end to end. On fine, loose sand, change of position becomes difficult. The grub now bends itself into a bow; it wriggles upon its back, upon its side; it crawls a little way; it digs and heaves with its mandibles. But let a less crumbling support come to its assistance; and pilgrimages of some length are not beyond its powers.

I reared my guests in a box divided into compartments by means of paper partitions. Each space, representing about the capacity of a Tachytes-cell, received its layer of sand, its pile of Mantes and its larva. And more than one disturbance arose in this refectory, where I had reckoned upon keeping the banqueters isolated one from the other, each at its special table. This larva, which had finished its ration the day before, was discovered next day in another chamber, where it was sharing its neighbour's repast. It had therefore climbed the partition, which for that matter was of no great height, or else had forced its way through some chink. This is enough, I think, to prove that the grub is not a strict stay-at-home, as are the larvæ of the Sitares and the Oil-beetles when devouring the ration of the Anthophora.

I imagine that, in the burrows of the Tachytes, the grub, when its heap of Mantes is consumed, moves from cell to cell until it has satisfied its appetite. Its subterranean excursions cannot cover a wide range, but they enable it to visit a few adjacent cells. I have mentioned how greatly the Tachytes' provision of Mantes varies.[4] The smaller rations certainly fall to the males, which are puny dwarfs compared with their companions; the more plentiful fall to the females. The parasitic grub to which fate has allotted the scanty masculine ration has not perhaps sufficient with this share; it wants an extra portion, which it can obtain by changing its cell. If it be favoured by chance, it will eat according to the measure of its hunger and will attain the full development of which its race allows; if it wander about without finding anything, it will fast and will remain small. This would explain the differences which I note in both the grubs and the pseudochrysalids, differences amounting in linear dimensions to a hundred per cent and more. The rations, rare or abundant according to the cells lit upon, would determine the size of the parasite.

[4] The essay on the Tachytes has not yet appeared in English. It will form part of a volume entitled *More Hunting Wasps.—Translator's Note.*

During the active period, the larva undergoes a few moults; I have witnessed at least one of these. The creature stripped of its skin appears as it was before, without any change of form. It instantly resumes its meal, which was interrupted while the old skin was shed; it embraces with its legs another Mantis on the heap and proceeds to nibble her. Whether simple or multiple, this moult has nothing in common with the renewals due to the hypermetamorphosis, which so profoundly change the creature's appearance.

Ten days' rearing in the partitioned box is enough to prove how right I was when I looked upon the parasitic larva feeding on Mantes as the origin of the pseudochrysalis, the object of my eager attention. The creature, which I kept supplied with additional food as long as it accepted it, stops eating at last. It becomes motionless, retracts its head slightly and bends itself into a hook. Then the skin splits across the head and down the thorax. The tattered slough is thrust back; and the pseudochrysalis appears in sight, absolutely naked. It is white at first, as the larva was; but by degrees and fairly rapidly it turns to the russet hue of virgin wax, with a brighter red at the tips of the various tubercles which indicate the future legs and mouthparts. This shedding of the skin, which leaves the body of the pseudochrysalis uncovered, recalls the mode of transformation observed in the Oil-beetles and is different from that of the Sitares and the Zonites, whose pseudochrysalis remains wholly enveloped in the skin of the secondary larva, a sort of bag which is sometimes loose, sometimes tight and always unbroken.

The mist that surrounded us at the outset is dispelled. This is indeed a Meloid, a true Meloid, one of the strangest anomalies among the parasites of its tribe. Instead of living on the honey of a Bee, it feeds on the skewerful of Mantes provided by a Tachytes. The North-American naturalists have taught us lately that honey is not always the diet of the Blister-beetles: some Meloidæ in the United States devour the packets of eggs laid by the Grasshoppers. This is a legitimate acquisition on their part, not an illegal seizure of the food-stores of others. No one, as far as I am aware, had as yet suspected the true parasitism of a carnivorous Meloid. It is nevertheless very remarkable to find in the Blister-beetles, on both sides of the Atlantic, this weakness for the flavour of Locust: one devours her eggs; the other a representative of the order, in the shape of the Praying Mantis and her kin.

Who will explain to me this predilection for the Orthopteron in a tribe whose chief, the Oil-beetle, accepts nothing but the mess of honey? Why do insects which appear close together in all our classifications possess such opposite tastes? If they spring from a common stock, how did the consumption of flesh supplant the consumption of honey? How did the Lamb become a Wolf? This is the great problem which was once set us, in an inverse form, by the Spotted Sapyga, a honey-eating relative of the flesh-eating Scolia.[5] I submit the question to whom it may concern.

[5] The essays on these will appear in the volume, entitled *The Hunting Wasps*, aforementioned.—*Translator's Note*.

The following year, at the beginning of June, some of my pseudochrysalids split open transversely behind the head and lengthwise down the whole of the median line of the back, except the last two or three segments. From it emerges the tertiary larva, which, from a simple examination with the pocket-lens, appears to me, in its general features, identical with the secondary larva, the one which eats the Tachytes' provisions. It is naked and pale-yellow, the colour of butter. It is active and wriggles with awkward movements. Ordinarily it lies upon its side, but it can also stand in the normal position. The creature is then trying to use its legs, without finding sufficient purchase to enable it to walk. A few days later, it relapses into complete repose.

Thirteen segments, including the head, which is large, with a quadrilateral cranium, rounded at the sides. Short antennæ, consisting of three knotted joints. Powerful curved mandibles, with two or three little teeth at the end, of a fairly bright red. Labial palpi rather bulky, short and with three joints, like the antennæ. The mouth-parts, labrum, mandibles and palpi are movable and stir slightly, as though seeking food. A small brown speck near the base of each antenna, marking the place of the future eyes. Prothorax wider than the segments that come after it. These are all of one width and are distinctly divided by a furrow and a slight lateral rim. Legs short, transparent, without a terminal claw. They are three-jointed stumps. Pale stigmata, eight pairs of them, placed as in the pseudochrysalis, that is, the first and largest pair on the line dividing the first two segments of the thorax and the seven others on the first seven abdominal segments. The secondary larva and the pseudochrysalis also have a very small stigma on the penultimate segment of the abdomen. This stigma has disappeared in the tertiary larva; at least I cannot detect it with the aid of a good magnifying-glass.

Lastly, we find the same strong mandibles as in the secondary larva, the same feeble legs, the same appearance of a Weevil-grub. The movements return, but are less clearly marked than in the primary form. The passage through the pseudochrysalid state has led to no change that is really worth describing. The creature, after this singular phase, is what it was before. The Meloes and Sitares, for that matter, behave similarly.

Then what can be the meaning of this pseudochrysalid stage, which, when passed, leads precisely to the point of departure? The Meloid seems to be revolving in a circle: it undoes what it has just done, it draws back after advancing. The idea sometimes occurs to me to look upon the pseudochrysalis as a sort of egg of a superior organization, starting from which the insect follows the ordinary law of entomological phases and passes through the successive stages of larva, nymph and perfect insect. The first hatching, that of the normal egg, makes the Meloid go through the larval dimorphism of the Anthrax and the Leucospis. The primary larva finds its way to the victuals; the secondary larva consumes them. The second hatching, that of the pseudochrysalis, reverts to the usual course, so that the insect passes through the three customary forms: larva, nymph, adult.

The tertiary larval stage is of brief duration, lasting about a fortnight. The larva then sheds its skin by a longitudinal rent along the back, as did the secondary larva, uncovering the nymph, in which we recognize the Beetle, the genus and species being almost determinable by the antennæ.

The second year's development turned out badly. The few nymphs which I obtained about the middle of June shrivelled up without attaining the perfect form. Some pseudochrysalids remained on my hands without showing any sign of approaching transformation. I attributed this delay to lack of warmth. I was in fact keeping them in the shade, on a what-not, in my study, whereas under natural conditions they are exposed to the hottest sun, beneath a layer of sand a few inches deep. To imitate these conditions without burying my charges, whose progress I wished to follow comfortably, I placed the pseudochrysalids that remained on a layer of fresh sand at the bottom of a glass receiver. Direct exposure to the sun was impracticable: it would have been fatal at a period when life is subterranean. To avoid it, I tied over the mouth of the receiver a few thicknesses of black cloth, to represent the natural screen of sand; and the apparatus thus prepared was exposed for some weeks to the most brilliant sunshine in my window. Under the cloth cover, which, owing to its colour, favours the absorption of heat, the temperature, during the day-time, became that of an oven; and yet the pseudochrysalids persisted in remaining stationary. The end of July was near and nothing indicated a speedy hatching. Convinced that my attempts at heating would be fruitless, I replaced the pseudochrysalids in the shade, on the shelves, in glass tubes. Here they passed a second year, still in the same condition.

June returned once more and with it the appearance of the tertiary larva, followed by the nymph. For the second time this stage of development was not exceeded; the one and only nymph that I succeeded in obtaining shrivelled, like those of the year before. Will these two failures, arising no doubt from the overdry atmosphere of my receivers, conceal from us the genus and the species of the Mantis-eating Meloid? Fortunately, no. The riddle is easily solved by deduction and comparison.

The only Melodiæ in my part of the country which, though their habits are still unknown, might correspond in size with either the larva or the pseudochrysalis in question are the Twelve-pointed Mylabris and Schaeffer's Cerocoma. I find the first in July on the flowers of the sea scabious; I find the second at the end of May and in June on the heads of the Îles d'Hyères everlasting. This last date is best-suited to explain the presence of the parasitic larva and its pseudochrysalis in the Tachytes' burrows from July onwards. Moreover, the Cerocoma is very abundant in the neighbourhood of the sand-heaps haunted by the Tachytes, while the Mylabris does not occur there. Nor is this all: the few nymphs obtained have curious antennæ, ending in a full, irregular tuft, the like of which is found only in the antennæ of the male Cerocoma. The Mylabris, therefore, must be eliminated; the antennæ, in the nymph, must be regularly jointed, as they are in the perfect insect. There remains the Cerocoma.

Any lingering doubts may be dispelled: by good fortune, a friend of mine, Dr. Beauregard, who is preparing a masterly work upon the Blister-beetles, had some pseudochrysalids of Schreber's Cerocoma in his possession. Having visited Sérignan for the purpose of scientific investigations, he had searched the Tachytes' sand-heaps in my company and taken back to Paris a few pseudochrysalids of grubs fed on Mantes, in order to follow their development. His attempts, like mine, had miscarried; but, on comparing the Sérignan pseudochrysalids with those of Schreber's Cerocoma, which came from Aramon, near Avignon, he was able to establish the closest resemblance between the two organisms. Everything therefore confirms the supposition that my discovery can relate only to Schaeffer's Cerocoma. As for the other, it must be eliminated: its extreme rarity in my neighbourhood is a sufficient reason.

It is tiresome that the diet of the Aramon Meloid is not known. If I allowed myself to be guided by analogy, I should be inclined to regard Schreber's Cerocoma as a parasite of *Tachytes tarsina*, who buries her hoards of young Locusts in the high sandy banks. In that case, the two Cerocomæ would have a similar diet. But I leave it to Dr. Beauregard to elucidate this important characteristic.

The riddle is deciphered: the Meloid that eats Praying Mantes is Schaeffer's Cerocoma, of whom I find plenty, in the spring, on the blossoms of the everlasting. Whenever I see it, my attention is attracted by an unusual peculiarity: the great difference of size that is able to exist between one specimen and another, albeit of the same sex. I see stunted creatures, females as well as males, which are barely one third the length of their better-developed companions. The Twelve-spotted Mylabris and the Four-spotted Mylabris present differences quite as pronounced in this respect.

The cause which makes a dwarf or a giant of the same insect, irrespective of its sex, can be only the smaller or greater quantity of food. If the larva, as I suspect, is obliged to find the Tachytes' game-larder for itself and to visit a second and a third, when the first is too frugally furnished, it may be imagined that the hazard of the road does not favour all in the same way, but rather allots abundance to one and penury to another. The grub that does not eat its fill remains small, while the one that gluts itself grows fat. These differences of size, in themselves, betray parasitism. If a mother's pains had amassed the food, or if the family had had the industry to obtain it direct instead of robbing others, the ration would be practically equal for all; and the inequalities in size would be reduced to those which often occur between the two sexes.

They speak, moreover, of a precarious, risky parasitism, wherein the Meloid is not sure of finding its food, which the Sitaris finds so deftly, getting itself carried by the Anthophora, after being born at the very entrance to the Bee's galleries and leaving its retreat only to slip into its host's fleece. A vagabond obliged to find for itself the food that suits it, the Cerocoma incurs the risk of Lenten fare.

One chapter is lacking to complete the history of Schaeffer's Cerocoma: that which treats of the beginning, the laying of the eggs, the egg itself and the primary larva. While watching the development of the Mantis-eating parasite, I took my precautions, in the first year, to discover its starting-point. By eliminating what was known to me and seeking among the Meloidæ of my neighbourhood for the size that corresponded with the pseudochrysalids unearthed from the Tachytes' burrows, I found, as I have said, only Schaeffer's Cerocoma and the Twelve-spotted Mylabris. I undertook to rear these in order to obtain their eggs.

As a standard of comparison, the Four-spotted Mylabris, of a more imposing size, was added to the first two. A fourth, *Zonitis mutica*, whom I did not need to consult, knowing that she was not connected with the matter in hand and being familiar with her pseudochrysalis, completed my school of egg-layers. I proposed, if possible, to obtain her primary larva. Lastly, I had formerly reared some Cantharides with the object of observing their egg-laying. In all, five species of Blister-beetles, reared in a breeding-cage, have left a few lines of notes in my records.

The method of rearing is of the simplest. Each species is placed under a large wire-gauze dome standing in a basin filled with earth. In the middle of the enclosure is a bottle full of water, in which the food soaks and keeps fresh. For the Cantharides, this is a bundle of ash-twigs; for the Four-spotted Mylabris, a bunch of bindweed (*Convolvus arvensis*) or psoralea (*P. biluminosa*), of which the insect nibbles only the corollæ. For the Twelve-spotted Mylabris, I provide blossoms of the scabious (*Scabiosa maritima*); for the Zonitis, the full-blown heads of the eryngo (*Eryngium campestre*); for Schaeffer's Cerocoma, the heads of the Îles d'Hyères everlasting (*Helichrysum stoechas*). These three last nibble more particularly the anthers, more rarely the petals, never the leaves.

A sorry intellect and sorry manners, which hardly repay the minute cares involved in the rearing. To browse, to love her lord, to dig a hole in the earth and carelessly to bury her eggs in it: that is the whole life of the adult Meloid. The dull creature acquires a little interest only at the moment when the male begins to toy with his mate. Every species has its own ritual in declaring its passion; and it is not beneath the dignity of the observer to witness the manifestations, sometimes so very strange, of the universal Eros, who rules the world and brings a tremor to even the lowest of the brute creation. This is the ultimate aim of the insect, which becomes transfigured for this solemn function and then dies, having no more to do.

A curious book might be written on the subject of love among the beasts. Long ago the subject tempted me. For a quarter of a century my notes have been slumbering, dustily, in a corner of my library. I extract from them the following details concerning the Cantharides. I am not the first, I know, to describe the amorous preludes of the Meloid of the Ash-tree; but the change of narrator may give the narrative a certain value: it confirms what has already been said and throws light upon some points which may have escaped notice.

A female Cantharides is peacefully nibbling her leaf. A lover comes upon the scene, approaches her from behind, suddenly mounts upon her back and embraces her with his two pairs of hind-legs. Then with his abdomen, which he lengthens as much as possible, he energetically slaps that of the female, on the right side and the left by turns. It is like the strokes of a washerwoman's bat, delivered with frenzied rapidity. With his antennæ and his fore-legs, which remain free, he furiously lashes the neck of the victim. While the blows fall thick as hail, in front and behind, the head and corselet of the amorous swain are shaken by an extravagant swaying and trembling. You would think that the creature was having an epileptic fit.

Meanwhile, the beloved makes herself small, opening her wing-cases slightly, hiding her head and tucking her abdomen under her, as though to escape the erotic thunderstorm that is bursting upon her back. But the paroxysm calms down. The male extends his fore-legs, shaken by a nervous tremor, like the arms of a cross and in this ecstatic posture seems to call upon the heavens to witness the ardour of his desires. The antennæ and the belly are held motionless, in a straight line; the head and the corselet alone continue to heave rapidly up and down. This period of repose does not last long. Short as it is, the female, her appetite undisturbed by the passionate protestations of her wooer, imperturbably resumes the nibbling of her leaf.

Another paroxysm bursts forth. Once more the male's blows rain upon the neck of the tightly-clasped victim, who hastens to bow her head upon her breast. But he has no intention of allowing his lady-love to escape. With his fore-legs, using a special notch placed at the juncture of the leg and the tarsus, he seizes both her antennæ. The tarsus folds back; and the antennæ are held as in a vice. The suitor pulls; and the callous one is forced to raise her head. In this posture the male reminds one of a horseman proudly sitting his steed and holding the reins in both hands. Thus mastering his mount, he is sometimes motionless and sometimes frenzied in his demonstrations. Then, with his long abdomen, he lashes the female's hinder-parts, first on one side, then on the other; the front part he flogs, hammers and pounds with blows of his antennæ, head and feet. The object of his desires will be unfeeling indeed if she refuse to surrender to so passionate a declaration.

Nevertheless she still requires entreating. The impassioned lover resumes his ecstatic immobility, with his quivering arms outstretched like the limbs of a cross. At brief intervals the amorous outbursts, with blows conscientiously distributed, recur in alternation with periods of repose, during which the male holds his fore-legs crosswise, or else masters the female by the bridle of her antennæ. At last the flagellated beauty allows herself to be touched by the charm attendant on his thumps. She yields. Coupling takes place and lasts for twenty hours. The heroic part of the male's performance is over. Dragged backwards behind the female, the poor fellow strives to uncouple himself. His mate carts him about from leaf to leaf, wherever she pleases, so that she may choose the bit of green stuff to her taste. Sometimes he also takes a gallant resolve and, like the female, begins to browse. You lucky creatures, who, so as not to lose a moment of your four or five weeks' existence, yoke together the cravings of love and hunger! Your motto is, "A short life and a merry one."

The Cerocoma, who is a golden green like the Cantharides, seems to have partly adopted the amorous rites of her rival in dress. The male, always the elegant sex in the insect tribe, wears special ornaments. The horns or antennæ, magnificently complicated, form as it were two tufts of a thick head of hair. It is to this that the name Cerocoma refers: the creature crested with its horns. When a bright sun shines into the breeding-cage, it is not long before the insects form couples on the bunch of everlastings. Hoisted on the female, whom he embraces and holds with his two pairs of hind-legs, the male sways his head and corselet up and down, all in a piece. This oscillatory movement has not the fiery precipitation of that of the Cantharides; it is calmer and as it were rhythmical. The abdomen moreover remains motionless and seems unskilled in those slaps, as of a washerwoman's bat, which the amorous denizen of the ash-tree so vigorously distributes with his belly.

While the front half of the body swings up and down, the fore-legs execute magnetic passes on either side of the tight-clasped female, moving with a sort of twirl, so rapidly that the eye can hardly follow them. The female appears insensible to this flagellatory twirl. She innocently curls her antennæ. The rejected suitor leaves her and moves on to another. His dizzy, twirling passes, his protestations are everywhere refused. The moment has not yet arrived, or rather the spot is not propitious. Captivity appears to weigh upon the future mothers. Before listening to their wooers they must have the open air, the sudden joyful flight from cluster to cluster on the sunlit slope, all gold with everlastings. Apart from the idyll of the twirling passes, a mitigated form of the Cantharides' blows, the Cerocoma refused to yield before my eyes to the last act of the bridal.

Among males the same oscillations of the body and the same lateral flagellations are frequently practised. While the upper one makes a tremendous to-do and whirls his legs, the one under him keeps quiet. Sometimes a third scatterbrain comes on the scene, sometimes even a fourth, and mounts upon the heap of his predecessors. The uppermost bobs up and down and makes swift rowing-strokes with his fore-legs; the others remain motionless. Thus are the sorrows of the rejected beguiled for a moment.

The Zonites, a rude clan, grazing on the heads of the prickly eryngo, despise all tender preliminaries. A few rapid vibrations of the antennæ on the males' part; and that is all. The declaration could not be briefer. The pairing, with the creatures placed end to end, lasts nearly an hour.

The Mylabres also must be very expeditious in their preliminaries, so much so that my cages, which were kept well-stocked for two summers, provided me with numerous batches of eggs without giving me a single opportunity of catching the males in the least bit of a flirtation. Let us therefore consider the egg-laying.

This takes place in August for our two species of Mylabres. In the vegetable mould which does duty as a floor to the wire-gauze dome, the mother digs a pit four-fifths of an inch deep and as wide as her body. This is the place for the eggs. The laying lasts barely half an hour. I have seen it last thirty-six hours with Sitares. This quickness of the Mylabris points to an incomparably less numerous family. The hiding-place is next closed. The mother sweeps up the rubbish with her fore-legs, collects it with the rake of her mandibles and pushes it back into the pit, into which she now descends to stamp upon the powdery layer and cram it down with her hind-legs, which I see swiftly working. When this layer is well packed, she starts raking together fresh material to complete the filling of the hole, which is carefully trampled stratum by stratum.

I take the mother from her pit while she is engaged in filling it up. Delicately, with the tip of a camel-hair pencil, I move her a couple of inches. The Beetle does not return to her batch of eggs, does not even look for it. She climbs up the wire gauze and proceeds to graze among her companions on the bindweed or scabious, without troubling herself further about her eggs, whose hiding-place is only half-filled. A second mother, whom I move only one inch, is no longer able to return to her task, or rather does not think of doing so. I take a third, after shifting her just as slightly, and, while the forgetful creature is climbing up the trellis-work, bring her back to the pit. I replace her with her head at the opening. The mother stands motionless, looking thoroughly perplexed. She sways her head, passes her front tarsi through her mandibles, then moves away and climbs to the top of the dome without attempting anything. In each of these three cases I have to finish filling in the pit myself. What then are this maternity, which the touch of a brush causes to forget its duties, and this memory, which is lost at a distance of an inch from the spot? Compare with these shortcomings of the adult the expert machinations of the primary larva, which knows where its victuals are and as its first action introduces itself into the dwelling of the host that is to feed it. How can time and experience be factors of instinct? The newborn animalcule amazes us with its foresight; the adult insect astonishes us with its stupidity.

With both Mylabres, the batch consists of some forty eggs, a very small number compared with those of the Oil-beetle and the Sitaris. This limited family was already foreseen, judging by the short space of time which the egg-layer spends in her underground lodging. The eggs of the Twelve-spotted Mylabris are white, cylindrical, rounded at both ends and measure a millimetre and a half in length by half a millimetre in width.[6] Those of the Four-spotted Mylabris are straw coloured and of an elongated oval, a trifle fuller at one end than at the other. Length, two millimetres; width, a little under one millimetre.[7]

[6] .058 x .019 inch. — *Translator's Note.*

[7] .078 x .039 inch. — *Translator's Note.*

Of all the batches of eggs collected, one alone hatched. The rest were probably sterile, a suspicion corroborated by the lack of pairing in the breeding-cage. Laid at the end of July, the eggs of the Twelve-spotted Mylabris began to hatch on the 5th of September. The primary larva of this Meloid is still unknown, so far as I am aware; and I shall describe it in detail. It will be the starting-point of a chapter which perhaps will give us some fresh sidelights upon the history of the hypermetamorphosis.

The larva is nearly 2 millimetres long.[8] Coming out of a good-sized egg, it is endowed with greater vigour than the larvæ of the Sitares and Oil-beetles. The head is large, rounded, slightly wider than the prothorax and of a rather brighter red. Mandibles powerful, sharp, curved, with the ends crossing, of the same colour as the head, darker at the tips. Eyes black, prominent, globular, very distinct. Antennæ fairly long, with three joints, the last thinner and pointed. Palpi very much pronounced.

[8] .078 inch. — *Translator's Note.*

The first thoracic segment has very nearly the same diameter as the head and is much longer than those which come after. It forms a sort of cuirass equal in length to almost three abdominal segments. It is squared off in front in a straight line and is rounded at the sides and at the back. Its colour is bright red. The second ring is hardly a third as long as the first. It is also red, but a little browner. The third is dark brown, with a touch of green to it. This tint is repeated throughout the abdomen, so that in the matter of colouring the creature is divided into two sections: the front, which is a fairly bright red, includes the head and the first two thoracic segments; the second, which is a greenish brown, includes the third thoracic segment and the nine abdominal rings.

The three pairs of legs are pale red, strong and long, considering the creature's smallness. They end in a single long, sharp claw.

The abdomen has nine segments, all of an olive brown. The membranous spaces which connect them are white, so that, from the second thoracic ring downwards, the tiny creature is alternatively ringed with white and olive brown. All the brown rings bristle with short, sparse hairs. The anal segment, which is narrower than the rest, bears at the tip two long cirri, very fine, slightly waved and almost as long as the whole abdomen.

This description enables us to picture a sturdy little creature, capable of biting lustily with its mandibles, exploring the country with its big eyes and moving about with six strong harpoons as a support. We no longer have to do with the puny louse of the Oil-beetle, which lies in ambush on a cichoriaceous blossom in order to slip into the fleece of a harvesting Bee; nor with the black atom of the Sitaris, which swarms in a heap on the spot where it is hatched, at the Anthophora's door. I see the young Mylabris striding eagerly up and down the glass tube in which it was born.

What is it seeking? What does it want? I give it a Bee, a Halictus,[9] to see if it will settle on the insect, as the Sitares and Oil-beetles would not fail to do. My offer is scorned. It is not a winged conveyance that my prisoners require.

[9] Cf. *Bramble-bees and Others:* chaps. xii. to xiv.—*Translator's Note.*

The primary larva of the Mylabris therefore does not imitate those of the Sitaris and the Oil-beetle; it does not settle in the fleece of its host to get itself carried to the cell crammed with victuals. The task of seeking and finding the heap of food falls upon its own shoulders. The small number of the eggs that constitute a batch also leads to the same conclusion. Remember that the primary larva of the Oil-beetle, for instance, settles on any insect that happens to pay a momentary visit to the flower in which the tiny creature is on the look-out. Whether this visitor be hairy or smooth-skinned, a manufacturer of honey, a canner of animal flesh or without any determined calling, whether she be Spider, Butterfly, Fly or Beetle makes no difference: the instant the little yellow louse espies the new arrival, it perches on her back and leaves with her. And now it all depends on luck! How many of these stray travellers must be lost; how many will never be carried into a warehouse full of honey, their sole food! Therefore, to remedy this enormous waste, the mother produces an innumerable family. The Oil-beetle's batch of eggs is prodigious. Prodigious too is that of the Sitaris, who is exposed to similar misadventures.

If, with her thirty or forty eggs, the Mylabris had to run the same risks, perhaps not one larva would reach the desired goal. For so strictly limited a family a safer method is needed. The young larva must not get itself carried to the game-basket, or more probably to the honey-pot, at the risk of never reaching it; it must travel on its own legs. Allowing myself to be guided by the logic of things, I shall therefore complete the story of the Twelve-spotted Mylabris as follows.

The mother lays her eggs underground near the spots frequented by the foster-mothers. The recently-hatched young grubs leave their lodgings in September and travel within a restricted radius in search of burrows containing food. The little creature's sturdy legs allow of these underground investigations. The mandibles, which are just as strong, necessarily play their part. The parasite, on forcing its way into the food-pit, finds itself faced with either the egg or the young larva of the Bee. These are competitors, whom it is important to get rid of as quickly as possible. The hooks of the mandibles now come into play, tearing the egg or the defenceless grub. After this act of brigandage, which may be compared with that of the primary larva of the Sitaris ripping open and drinking the contents of the Anthophora's egg, the Meloid, now the sole possessor of the victuals, doffs its battle array and becomes the pot-bellied grub, the consumer of the property so brutally acquired. These are merely suspicions on my part, nothing more. Direct observation will, I believe, confirm them, so close is their connection with the known facts.

Two Zonites, both visitors of the eryngo-heads during the heats of summer, are among the Meloidæ of my part of the country. They are *Zonitis mutica* and *Z. præusta*. I have spoken of the first in another volume;[10] I have mentioned its pseudochrysalis found in the cells of two Osmiæ, namely, the Three-pronged Osmia, which piles its cells in a dry bramble-stem, and the Three-horned Osmia and also Latreille's Osmia, both of which exploit the nests of the Chalicodoma of the Sheds. The second Zonitis is to-day adding its quota of evidence to a story which is still very incomplete. I have obtained the Burnt Zonitis, in the first place, from the cotton pouches of *Anthidium scapulare*, who, like the Three-toothed Osmia, makes her nests in the brambles; in the second place, from the wallets of *Megachile sericans*, made with little round disks of the leaves of the common acacia; in the third place, from the cells which *Anthidium bellicosum*[11] builds with partitions of resin in the shell of a dead Snail. This last Anthidium is the victim also of the Unarmed Zonitis. Thus we have two closely-related exploiters for the same victim.

[10] Cf. *Bramble-bees and Others:* chaps. i., iii. and x.—*Translator's Note.*

[11] For the Cotton-bee, Leaf-cutter and Resin-bee mentioned, cf. *Bramble-bees and Others: passim.*—*Translator's Note.*

During the last fortnight of July, I witness the emergence of the Burnt Zonitis from the pseudochrysalis. The latter is cylindrical, slightly curved and rounded at both ends. It is closely wrapped in the cast skin of the secondary larva, a skin consisting of a diaphanous bag, without any outlet, with running along each side a white tracheal thread which connects the various stigmatic apertures. I easily recognize the seven abdominal stigmata; they are round and diminish slightly in width from front to back. I also detect the thoracic stigma. Lastly, I perceive the legs, which are quite small, with weak claws, incapable of supporting the creature. Of the mouth-parts I see plainly only the mandibles, which are short, weak and brown. In short, the secondary larva was soft, white, big-bellied, blind, with rudimentary legs. Similar results were furnished by the shed skin of the secondary larva of *Zonitis mutica*, consisting, like the other, of a bag without an opening, fitting closely over the pseudochrysalis.

Let us continue our examination of the relics of the Burnt Zonitis. The pseudochrysalis is red, the colour of a cough-lozenge. It remains intact after opening, except in front, where the adult insect has emerged. In shape it is a cylindrical bag, with firm, elastic walls. The segmentation is plainly visible. The magnifying-glass shows the fine star-shaped dots already observed in the Unarmed Zonitis. The stigmatic apertures have a projecting, dark-red rim. They are all, even the last, clearly marked. The signs of the legs are mere studs, hardly protruding, a little darker than the rest of the skin. The cephalic mask is reduced to a few mouldings which are not easy to distinguish.

At the bottom of this pseudochrysalidal sheath I find a little white wad which, when placed in water, softened and then patiently unravelled with the tip of a paint-brush, yields a white, powdery substance, which is uric acid, the usual product of the work of the nymphosis, and a rumpled membrane, in which I recognize the cast skin of the nymph. There should still be the tertiary larva, of which I see not a trace. But, on taking a needle and gradually breaking the envelope of the pseudochrysalis, after soaking it awhile in water, I see it dividing into two layers, one an outer layer, brittle, horny in appearance and currant-red; the other an inner layer, consisting of a transparent, flexible pellicle. There can be no doubt that this inner layer represents the tertiary larva, whose skin is left adhering to the envelope of the pseudochrysalis. It is fairly thick and tough, but I cannot detach it except in shreds, so closely does it adhere to the horny, crumbly sheath.

Since I possessed a fair number of pseudochrysalids, I sacrificed a few in order to ascertain their contents on the approach of the final transformations. Well, I never found anything that I could detach; I never succeeded in extracting a larva in its tertiary form, though this larva is so easily obtained from the amber pouches of the Sitares and, in the Oil-beetles and Cerocomæ, emerges of its own accord from the split wrapper of the pseudochrysalis. When, for the first time, the stiff shell encloses a body which does not adhere to the rest, this body is a nymph and nothing else. The wall surrounding it is a dull white inside. I attribute this colouring to the cast skin of the tertiary larva, which was inseparably fixed to the shell of the pseudochrysalis.

The Zonites, therefore, display a peculiarity which is not offered by the other Meloidæ, namely, a series of tightly-fitting shells, one within the other. The pseudochrysalis is enclosed in the skin of the secondary larva, a skin which forms a pouch without an orifice, fitted very closely to its contents. The slough of the tertiary larva fits even more closely to the inner surface of the pseudochrysalid sheath. The nymph alone does not adhere to its envelope. In the Cerocomæ and the Oil-beetles, each form of the hypermetamorphosis becomes detached from the preceding skin by a complete extraction; the contents are removed from the ruptured container and have no further connection with it. In the Sitares, the successive casts are not ruptured and remain enclosed inside one another, but with an interval between, so that the tertiary larva can move and turn as it wishes in its multiple enclosure. In the Zonites, there is the same arrangement, with this difference, that, until the nymph appears, there is no empty space between one slough and the next. The tertiary larva cannot budge. It is not free, as witness its cast skin, which fits so precisely to the envelope of the pseudochrysalis. This form would therefore pass unperceived if its existence were not proclaimed by the membrane which lines the inside of the pseudochrysalid pouch.

To complete the story of the Zonites, the primary larva is lacking. I do not yet know it, for, when rearing the insect under wire-gauze covers, I never succeeded in obtaining a batch of eggs.

CHAPTER VII
THE CAPRICORN

My youthful meditations owe some happy moments to Condillac's[1] famous statue which, when endowed with the sense of smell, inhales the scent of a rose and out of that single impression creates a whole world of ideas. My twenty-year-old mind, full of faith in syllogisms, loved to follow the deductive jugglery of the abbé-philosopher: I saw, or seemed to see, the statue take life in that action of the nostrils, acquiring attention, memory, judgment and all the psychological paraphernalia, even as still waters are aroused and rippled by the impact of a grain of sand. I recovered from my illusion under the instruction of my abler master, the animal. The Capricorn shall teach us that the problem is more obscure than the abbé led me to believe.

[1] Étienne Bonnot de Condillac, Abbé de Mureaux (1715-1780), the leading exponent of sensational philosophy. His most important work is the *Traité des sensations*, in which he imagines a statue, organized like a man, and endows it with the senses one by one, beginning with that of smell. He argues by a process of imaginative reconstruction that all human faculties and all human knowledge are merely transformed sensation, to the exclusion of any other principle, that, in short, everything has its source in sensation: man is nothing but what he has acquired. — *Translator's Note.*

When wedge and mallet are at work, preparing my provision of firewood under the grey sky that heralds winter, a favourite relaxation creates a welcome break in my daily output of prose. By my express orders, the woodman has selected the oldest and most ravaged trunks in his stack. My tastes bring a smile to his lips; he wonders by what whimsy I prefer wood that is worm-eaten, *chirouna*, as he calls it, to sound wood, which burns so much better. I have my views on the subject; and the worthy man submits to them.

And now to us two, O my fine oak-trunk seamed with scars, gashed with wounds whence trickle the brown drops smelling of the tanyard. The mallet drives home, the wedges bite, the wood splits. What do your flanks contain? Real treasures for my studies. In the dry and hollow parts, groups of various insects, capable of living through the bad season of the year, have taken up their winter quarters: in the low-roofed galleries, galleries built by some Buprestis Beetle, Osmiæ, working their paste of masticated leaves, have piled their cells one above the other; in the deserted chambers and vestibules, Megachiles have arranged their leafy jars; in the live wood, filled with juicy saps, the larvæ of the Capricorn (*Cerambyx miles*), the chief author of the oak's undoing, have set up their home.

Strange creatures, of a verity, are these grubs, for an insect of superior organization: bits of intestines crawling about! At this time of year, the middle of autumn, I meet them of two different ages. The older are almost as thick as one's finger; the others hardly attain the diameter of a pencil. I find, in addition, pupæ more or less fully coloured, perfect insects, with a distended abdomen, ready to leave the trunk when the hot weather comes again. Life inside the wood, therefore, lasts three years. How is this long period of solitude and captivity spent? In wandering lazily through the thickness of the oak, in making roads whose rubbish serves as food. The horse in Job swallows the ground[2] in a figure of speech; the Capricorn's grub eats its way literally. With its carpenter's-gouge, a strong black mandible, short, devoid of notches, scooped into a sharp-edged spoon, it digs the opening of its tunnel. The piece cut out is a mouthful which, as it enters the stomach, yields its scanty juices and accumulates behind the worker in heaps of wormed wood. The refuse leaves room in front by passing through the worker. A labour at once of nutrition and of road-making, the path is devoured while constructed; it is blocked behind as it makes way ahead. That, however, is how all the borers who look to wood for victuals and lodging set about their business.

[2] "Chafing and raging, he swalloweth the ground, neither doth he make account when the noise of the trumpet soundeth."—Job, xxxix, 23 (Douai version).—*Translator's Note.*

For the harsh work of its two gouges, or curved chisels, the larva of the Capricorn concentrates its muscular strength in the front of its body, which swells into a pestle-head. The Buprestis-grubs, those other industrious carpenters, adopt a similar form; they even exaggerate their pestle. The part that toils and carves hard wood requires a robust structure; the rest of the body, which has but to follow after, continues slim. The essential thing is that the implement of the jaws should possess a solid support and a powerful motor. The Cerambyx-larva strengthens its chisels with a stout, black, horny armour that surrounds the mouth; yet, apart from its skull and its equipment of tools, the grub has a skin as fine as satin and as white as ivory. This dead white comes from a copious layer of grease which the animal's spare diet would not lead us to suspect. True, it has nothing to do, at every hour of the day and night, but gnaw. The quantity of wood that passes into its stomach makes up for the dearth of nourishing elements.

The legs, consisting of three pieces, the first globular, the last sharp-pointed, are mere rudiments, vestiges. They are hardly a millimetre[3] long. For this reason, they are of no use whatever for walking; they do not even bear upon the supporting surface, being kept off it by the obesity of the chest. The organs of locomotion are something altogether different. The Cetonia-grub[4] has shown us how, with the aid of the hairs and the pad-like excrescences upon its spine, it manages to reverse the universally-accepted usage and to wriggle along on its back. The grub of the Capricorn is even more ingenious: it moves at the same time on its back and belly; instead of the useless legs of the thorax, it has a walking-apparatus almost resembling feet, which appear, contrary to every rule, on the dorsal surface.

[3] .039 inch. — *Translator's Note.*

[4] For the grub of the Cetonia, or Rose-chafer, cf. *The Life and Love of the Insect*, by J. Henri Fabre, translated by Alexander Teixeira de Mattos: chap. xi. — *Translator's Note.*

The first seven segments of the abdomen have, both above and below, a four-sided facet, bristling with rough protuberances. This the grub can either expand or contract, making it stick out or lie flat at will. The upper facets consist of two excrescences separated by the mid-dorsal line; the lower ones have not this divided appearance. These are the organs of locomotion, the ambulacra. When the larva wishes to move forwards, it expands its hinder ambulacra, those on the back as well as those on the belly, and contracts its front ones. Fixed to the side of the narrow gallery by their ridges, the hind-pads give the grub a purchase. The flattening of the fore-pads, by decreasing the diameter, allows it to slip forward and to take half a step. To complete the step, the hind-quarters have to be brought up the same distance. With this object, the front pads fill out and provide support, while those behind shrink and leave free scope for their segments to contract.

With the double support of its back and belly, with alternate puffings and shrinkings, the animal easily advances or retreats along its gallery, a sort of mould which the contents fill without a gap. But, if the locomotory pads grip only on one side, progress becomes impossible. When placed on the smooth wood of my table, the animal wriggles slowly; it lengthens and shortens without advancing by a hair's-breadth. Laid on the surface of a piece of split oak, a rough, uneven surface, due to the gash made by the wedge, it twists and writhes, moves the front part of its body very slowly from left to right and right to left, lifts it a little, lowers it and begins again. These are the most extensive movements made. The vestigial legs remain inert and absolutely useless.

Then why are they there? Better to lose them altogether, if it be true that crawling inside the oak has deprived the animal of the good legs with which it started. The influence of environment, so well-inspired in endowing the grub with ambulatory pads, becomes a mockery when it leaves it these ridiculous stumps. Can the structure, perchance, be obeying other rules than those of environment?

Though the useless legs, the germs of the future limbs, persist, there is no sign in the grub of the eyes wherewith the Cerambyx will be richly gifted. The larva has not the least trace of organs of vision. What would it do with sight, in the murky thickness of a tree-trunk?

Hearing is likewise absent. In the never-troubled silence of the oak's inmost heart, the sense of hearing would be a non-sense. Where sounds are lacking, of what use is the faculty of discerning them? Should there be any doubts, I will reply to them with the following experiment. Split lengthwise, the grub's abode leaves a half-tunnel wherein I can watch the occupant's doings. When left alone, it now gnaws the front of its gallery, now rests, fixed by its ambulacra to the two sides of the channel. I avail myself of these moments of quiet to enquire into its power of perceiving sounds. The banging of hard bodies, the ring of metallic objects, the grating of a file upon a saw are tried in vain. The animal remains impassive. Not a wince, not a move of the skin; no sign of awakened attention. I succeed no better when I scratch the wood close by with a hard point, to imitate the sound of some neighbouring larva gnawing the intervening thickness. The indifference to my noisy tricks could be no greater in a lifeless object. The animal is deaf.

Can it smell? Everything tells us no. Scent is of assistance in the search for food. But the Capricorn-grub need not go in quest of eatables: it feeds on its home, it lives on the wood that gives it shelter. Let us make an attempt or two, however. I scoop in a log of fresh cypress-wood a groove of the same diameter as that of the natural galleries and I place the worm inside it. Cypress-wood is strongly-scented; it possesses in a high degree that resinous aroma which characterizes most of the pine family. Well, when laid in the odoriferous channel, the larva goes to the end, as far as it can go, and makes no further movement. Does not this placid quiescence point to the absence of a sense of smell? The resinous flavour, so strange to the grub which has always lived in oak, ought to vex it, to trouble it; and the disagreeable impression ought to be revealed by a certain commotion, by certain attempts to get away. Well, nothing of the kind happens: once the larva has found the right position in the groove, it does not stir. I do more: I set before it, at a very short distance, in its normal canal, a piece of camphor. Again, no effect. Camphor is followed by naphthaline. Still nothing. After these fruitless endeavours, I do not think that I am going too far when I deny the creature a sense of smell.

Taste is there, no doubt. But such taste! The food is without variety: oak, for three years at a stretch, and nothing else. What can the grub's palate appreciate in this monotonous fare? The tannic relish of a fresh piece, oozing with sap; the uninteresting flavour of an over-dry piece, robbed of its natural condiment: these probably represent the whole gustative scale.

There remains touch, the far-spreading passive sense common to all live flesh that quivers under the goad of pain. The sensitive schedule of the Cerambyx-grub, therefore, is limited to taste and touch, both exceedingly obtuse. This almost brings us to Condillac's statue. The imaginary being of the philosopher had one sense only, that of smell, equal in delicacy to our own; the real being, the ravager of the oak, has two, inferior, even when put together, to the former, which so plainly perceived the scent of a rose and distinguished it so clearly from any other. The real case will bear comparison with the fictitious.

What can be the psychology of a creature possessing such a powerful digestive organism combined with such a feeble set of senses? A vain wish has often come to me in my dreams: it is to be able to think, for a few minutes, with the crude brain of my Dog, to see the world with the faceted eyes of a Gnat. How things would change in appearance! They would change much more if interpreted by the intellect of the grub. What have the lessons of touch and taste contributed to that rudimentary receptacle of impressions? Very little; almost nothing. The animal knows that the best bits possess an astringent flavour; that the sides of a passage not carefully planed are painful to the skin. This is the utmost limit of its acquired wisdom. In comparison, the statue with the sensitive nostrils was a marvel of knowledge, a paragon too generously endowed by its inventor. It remembered, compared, judged, reasoned: does the drowsy, digesting paunch remember? Does it compare? Does it reason? I defined the Capricorn-grub as a bit of an intestine that crawls about. The undeniable accuracy of this definition provides me with my answer: the grub has the aggregate of sense-impressions that a bit of an intestine may hope to have.

And this nothing-at-all is capable of marvellous acts of foresight; this belly, which knows hardly anything of the present, sees very clearly into the future. Let us take an illustration on this curious subject. For three years on end, the larva wanders about in the thick of the trunk; it goes up, goes down, turns to this side and that; it leaves one vein for another of better flavour, but without moving too far from the inner depths, where the temperature is milder and greater safety reigns. A day is at hand, a dangerous day for the recluse obliged to quit its excellent retreat and face the perils of the surface. Eating is not everything: we have to get out of this. The larva, so well-equipped with tools and muscular strength, finds no difficulty in going where it pleases, by boring through the wood; but does the coming Capricorn, whose short spell of life must be spent in the open air, possess the same advantages? Hatched inside the trunk, will the long-horned Beetle be able to clear itself a way of escape?

That is the difficulty which the worm solves by inspiration. Less versed in things of the future, despite my gleams of reason, I resort to experiment with a view to fathoming the question. I begin by ascertaining that the Capricorn, when he wishes to leave the trunk, is absolutely unable to make use of the tunnel wrought by the larva. It is a very long and very irregular maze, blocked with great heaps of wormed wood. Its diameter decreases progressively from the final blind alley to the starting-point. The larva entered the timber as slim as a tiny bit of straw; it is to-day as thick as one's finger. In its three years' wanderings, it always dug its gallery according to the mould of its body. Evidently, the road by which the larva entered and moved about cannot be the Capricorn's exit-way: his immoderate antennæ, his long legs, his inflexible armour-plates would encounter an insuperable obstacle in the narrow, winding corridor, which would have to be cleared of its wormed wood and, moreover, greatly enlarged. It would be less fatiguing to attack the untouched timber and dig straight ahead. Is the insect capable of doing so? We shall see.

I make some chambers of suitable size in oak logs chopped in two; and each of my artificial cells receives a newly-transformed Cerambyx, such as my provisions of firewood supply, when split by the wedge, in October. The two pieces are then joined and kept together with a few bands of wire. June comes. I hear a scraping inside my billets. Will the Capricorns come out, or not? The delivery does not seem difficult to me: there is hardly three-quarters of an inch to pierce. Not one emerges. When all is silence, I open my apparatus. The captives, from first to last, are dead. A vestige of sawdust, less than a pinch of snuff, represents all their work.

I expected more from those sturdy tools, their mandibles. But, as we have seen before, the tool does not make the workman.[5] In spite of their boring-implements, the hermits die in my cases for lack of skill. I subject others to less arduous tests. I enclose them in spacious reed-stumps, equal in diameter to the natal cell. The obstacle to be pierced is the natural diaphragm, a yielding partition two or three millimetres[6] thick. Some free themselves; others cannot. The less valiant ones succumb, stopped by the frail barrier. What would it be if they had to pass through a thickness of oak?

[5] Cf. *The Life and Love of the Insect:* chap. iii. "The tool does not make the workman. The insect exerts its gifts as a specialist with any kind of tool wherewith it is supplied. It can saw with a plane or plane with a saw, like the model workman of whom Franklin tells us."—*Translator's Note.*

[6] .078 to .117 inch.—*Translator's Note.*

We are now persuaded: despite his stalwart appearance, the Capricorn is powerless to leave the tree-trunk by his unaided efforts. It therefore falls to the worm, to the wisdom of that bit of an intestine, to prepare the way for him. We see renewed, in another form, the feats of prowess of the Anthrax, whose pupa, armed with trepans, bores through rock on the feeble Fly's behalf. Urged by a presentiment that to us remains an unfathomable mystery, the Cerambyx-grub leaves the inside of the oak, its peaceful retreat, its unassailable stronghold, to wriggle towards the outside, where lives the foe, the Woodpecker, who may gobble up the succulent little sausage. At the risk of its life, it stubbornly digs and gnaws to the very bark, of which it leaves no more intact than the thinnest film, a slender screen. Sometimes, even, the rash one opens the window wide.

This is the Capricorn's doorway. The insect will have but to file the screen a little with its mandibles, to bump against it with its forehead, in order to bring it down; it will even have nothing to do when the window is free, as often happens. The unskilled carpenter, burdened with his extravagant head-dress, will emerge from the darkness through this opening when the summer heats arrive.

After the cares of the future come the cares of the present. The larva, which has just opened the aperture of escape, retreats some distance down its gallery and, in the side of the exit-way, digs itself a transformation-chamber more sumptuously furnished and barricaded than any that I have ever seen. It is a roomy niche, shaped like a flattened ellipsoid, the length of which reaches some eighty to a hundred millimetres.[7] The two axes of the cross-section vary: the horizontal measures twenty-five to thirty millimetres;[8] the vertical measures only fifteen.[9] This greater dimension of the cell, where the thickness of the perfect insect is concerned, leaves a certain scope for the action of its legs when the time comes for forcing the barricade, which is more than a close-fitting mummy-case would do.

[7] 3 to 4 inches. — *Translator's Note.*

[8] .975 to 1.17 inch. — *Translator's Note.*

[9] .585 inch. — *Translator's Note.*

The barricade in question, a door which the larva builds to exclude the dangers from without, is two- and even three-fold. Outside, it is a stack of woody refuse, of particles of chopped timber; inside, a mineral hatch, a concave cover, all in one piece, of a chalky white. Pretty often, but not always, there is added to these two layers an inner casing of shavings. Behind this compound door, the larva makes its arrangements for the metamorphosis. The sides of the chamber are rasped, thus providing a sort of down formed of ravelled woody fibres, broken into minute shreds. The velvety matter, as and when obtained, is applied to the wall in a continuous felt at least a millimetre thick.[10] The chamber is thus padded throughout with a fine swan's-down, a delicate precaution taken by the rough worm on behalf of the tender pupa.

[10] .039 inch. — *Translator's Note.*

Let us hark back to the most curious part of the furnishing, the mineral hatch or inner door of the entrance. It is an elliptical skull-cap, white and hard as chalk, smooth within and knotted without, resembling more or less closely an acorn-cup. The knots show that the matter is supplied in small, pasty mouthfuls, solidifying outside in slight projections which the animal does not remove, being unable to get at them, and polished on the inside surface, which is within the worm's reach. What can be the nature of that singular lid whereof the Cerambyx furnishes me with the first specimen? It is as hard and brittle as a flake of lime-stone. It can be dissolved cold in nitric acid, discharging little gaseous bubbles. The process of solution is a slow one, requiring several hours for a tiny fragment. Everything is dissolved, except a few yellowish flocks, which appear to be of an organic nature. As a matter of fact, a piece of the lid, when subjected to heat, blackens, which proves the presence of an organic glue cementing the mineral matter. The solution becomes muddy if oxalate of ammonia be added and deposits a copious white precipitate. These signs indicate calcium carbonate. I look for urate of ammonia, that constantly-recurring product of the various stages of the metamorphoses. It is not there: I find not the least trace of murexide. The lid, therefore, is composed solely of carbonate of lime and of an organic cement, no doubt of an albuminous character, which gives consistency to the chalky paste.

Had circumstances served me better, I should have tried to discover in which of the worm's organs the stony deposit dwells. I am, however, convinced: it is the stomach, the chylific ventricle, that supplies the chalk. It keeps it separate from the food, either as original matter or as a derivative of the ammonium urate; it purges it of all foreign bodies, when the larval period comes to an end, and holds it in reserve until the time comes to disgorge it. This freestone-factory causes me no astonishment: when the manufacturer undergoes his change, it serves for various chemical works. Certain Oil-beetles, such as the Sitaris, locate in it the urate of ammonia, the refuse of the transformed organism; the Sphex, the Pelopæi, the Scoliæ,[11] use it to manufacture the shellac wherewith the silk of the cocoon is varnished. Further investigations will only swell the aggregate of the products of this obliging organ.

[11] Three species of Digger-wasps.—*Translator's Note.*

When the exit-way is prepared and the cell upholstered in velvet and closed with a three-fold barricade, the industrious worm has concluded its task. It lays aside its tools, sheds its skin and becomes a nymph, a pupa, weakness personified, in swaddling-clothes, on a soft couch. The head is always turned towards the door. This is a trifling detail in appearance; but it is everything in reality. To lie this way or that in the long cell is a matter of great indifference to the worm, which is very supple, turning easily in its narrow lodging and adopting whatever position it pleases. The coming Capricorn will not enjoy the same privileges. Stiffly girt in his horn cuirass, he will not be able to turn from end to end; he will not even be capable of bending, if some sudden wind should make the passage difficult. He must absolutely find the door in front of him, lest he perish in the casket. Should the grub forget this little formality, should it lie down to its nymphal sleep with its head at the back of the cell, the Capricorn is infallibly lost: his cradle becomes a hopeless dungeon.

But there is no fear of this danger: the knowledge of the bit of an intestine is too sound in things of the future for the grub to neglect the formality of keeping its head to the door. At the end of spring, the Capricorn, now in possession of his full strength, dreams of the joys of the sun, of the festivals of light. He wants to get out. What does he find before him? A heap of filings easily dispersed with his claws; next, a stone lid which he need not even break into fragments: it comes undone in one piece; it is removed from its frame with a few pushes of the forehead, a few tugs of the claws. In fact, I find the lid intact on the threshold of the abandoned cells. Last comes a second mass of woody remnants as easy to disperse as the first. The road is now free: the Cerambyx has but to follow the spacious vestibule, which will lead him, without the possibility of mistake, to the exit. Should the window not be open, all that he has to do is to gnaw through a thin screen: an easy task; and behold him outside, his long antennæ aquiver with excitement.

What have we learnt from him? Nothing from him; much from his grub. This grub, so poor in sensory organs, gives us with its prescience no little food for reflection. It knows that the coming Beetle will not be able to cut himself a road through the oak and it bethinks itself of opening one for him at its own risk and peril. It knows that the Cerambyx, in his stiff armour, will never be able to turn and make for the orifice of the cell; and it takes care to fall into its nymphal sleep with its head to the door. It knows how soft the pupa's flesh will be and upholsters the bedroom with velvet. It knows that the enemy is likely to break in during the slow work of the transformation and, to set a bulwark against his attacks, it stores a calcium pap inside its stomach. It knows the future with a clear vision, or, to be accurate, behaves as though it knew the future. Whence did it derive the motives of its actions? Certainly not from the experience of the senses. What does it know of the outside world? Let us repeat, as much as a bit of an intestine can know. And this sense-less creature astounds us! I regret that the clever logician, instead of conceiving a statue smelling a rose, did not imagine it gifted with some instinct. How quickly he would have recognized that, quite apart from sense-impressions, the animal, including man, possesses certain psychological resources, certain inspirations that are innate and not acquired!

CHAPTER VIII
THE PROBLEM OF THE SIREX

The cherry-tree supports a small jet-black Capricorn, *Cerambyx cerdo*, whose larval habits it was as well to study in order to learn whether the instincts are modified when the form and the organization remain identical. Has this pigmy of the family the same talents as the giant, the ravager of the oak-tree? Does it work on the same principles? The resemblance between the two, both in the larval state and in that of the perfect insect, is complete; the denizen of the cherry-tree is an exact replica, on a smaller scale, of the denizen of the oak. If instinct is the inevitable consequence of the organism, we ought to find in the two insects a strict similarity of habits; if instinct is, on the other hand, a special aptitude favoured by the organs, we must expect variations in the industry exercised. For the second time the alternative is forced upon our attention: do the implements govern the practice of the craft, or does the craft govern the employment of the implements? Is instinct derived from the organ, or is the organ instinct's servant? An old dead cherry-tree will answer our question.

Beneath its ragged bark, which I lift in wide strips, swarms a population of larvæ all belonging to *Cerambyx cerdo*. There are big larvæ and little larvæ; moreover, they are accompanied by nymphs. These details tell us of three years of larval existence, a duration of life frequent in the Longicorn series. If we hunt the thick of the trunk, splitting it again and again, it does not show us a single grub anywhere; the entire population is encamped between the bark and the wood. Here we find an inextricable maze of winding galleries, crammed with packed sawdust, crossing, recrossing, shrinking into little alleys, expanding into wide spaces and cutting, on the one hand, into the surface layer of the sap-wood and, on the other, into the thin sheets of the inner bark. The position speaks for itself: the larva of the little Capricorn has other tastes than its large kinsman's; for three years it gnaws the outside of the trunk beneath the thin covering of the bark, while the other seeks a deeper refuge and gnaws the inside.

The dissimilarity is yet more marked in the preparations for the nymphosis. Then the worm of the cherry-tree leaves the surface and penetrates into the wood to a depth of about two inches, leaving behind it a wide passage, which is hidden on the outside by a remnant of bark that has been discreetly spared. This spacious vestibule is the future insect's path of release; this screen of bark, easily destroyed, is the curtain that masks the exit-door. In the heart of the wood the larva finally scoops out the chamber destined for the nymphosis. This is an egg-shaped recess an inch and a quarter to an inch and three-quarters in length by two-fifths of an inch in diameter. The walls are bare, that is to say, they are not lined with the blanket of shredded fibres dear to the Capricorn of the Oak. The entrance is blocked first by a plug of fibrous sawdust, then by a chalky lid, similar, except in point of size, to that with which we are already familiar. A thick layer of fine sawdust packed into the concavity of the chalky lid, completes the barricade. Need I add that the grub lies down and goes to sleep, for the nymphosis, with its head against the door? Not one forgets to take this precaution.

The two Capricorns have, in short, the same system of closing their cells. Note above all the lens-shaped stony lid. In each case we find the same chemical composition, the same formation, like the cup of an acorn. Dimensions apart, the two structures are identical. But no other genus of Longicorn, so far as I am aware, practises this craft. I will therefore complete the classic description of the Cerambyx-beetles by adding one characteristic: they seal their metamorphosis-chambers with a chalk slab.

The similarities of habit go no farther, despite the identity of structure. There is even a very sharp contrast between the methods pursued. The Capricorn of the Oak inhabits the deep layers of the trunk; the Capricorn of the Cherry-tree inhabits the surface. In the preparations for the transformation, the first ascends from the wood to the bark, the second descends from the bark to the wood; the first risks the perils of the outer world, the second shuns them and seeks a retreat inside. The first hangs the walls of its chamber with velvet, the second knows nothing of this luxury. Though the work is almost the same in its results, it is at least carried out by contrary methods.

The tool, therefore, does not govern the trade. This is what the two Cerambyx-beetles tell us.

Let us vary the testimony of the Longicorns. I am not selecting; I am recording it in the order of my discoveries. The Shagreen Saperda (*S. carcharias*) lives in the black poplar; the Scalary Saperda (*S. scalaris*) lives in the cherry-tree. In both we find the same organization and the same implements, as is fitting in two closely-related species. The Saperda of the Poplar adopts the method of the Capricorn of the Oak in its general features. It inhabits the interior of the trunk. On the approach of the transformation, it makes an exit-gallery, the door of which is open or else masked by a remnant of bark. Then, retracing its steps, it blocks the passage with a barricade of coarse packed shavings; and, at a depth of about eight inches, not far from the heart of the tree, it hollows out a cavity for the nymphosis without any particular upholstering. The defensive system is limited to the long column of shavings. To deliver itself, the insect will only have to push the heap of woody rubbish back, in so many lots; the path will open in front of it ready-made. If some screen of bark hide the gallery from the outside, its mandibles will easily dispose of that: it is soft and not very thick.

The Scalary Saperda imitates the habits of its messmate, the Capricorn of the Cherry-tree. Its larva lives between the wood and the bark. To undergo its transformation, it goes down instead of coming up. In the sap-wood, parallel with the surface of the trunk, under a layer of wood barely a twenty-fifth of an inch in thickness, it makes a cylindrical cell, rounded at the ends and roughly padded with ligneous fibres. A solid plug of shavings barricades the entrance, which is not preceded by any vestibule. Here the work of deliverance is the simplest. The Saperda has only to clear the door of his chamber to find beneath his mandibles the little bit of bark that remains to be pierced. As you see, we once more have to do with two specialists, each working in his own manner with the same tools.

The Buprestes, as zealous as the Longicorns in the destruction of trees, whether sound or ailing, tell us the same tale as the Cerambyx- and Saperda-beetles. The Bronze Buprestis (*B. ænea*) is an inmate of the black poplar. Her larva gnaws the interior of the trunk. For the nymphosis it installs itself near the surface in a flattened, oval cell, which is prolonged at the back by the wandering-gallery, firmly packed with wormed wood, and in front by a short, slightly curved vestibule. A layer of wood not a twenty-fifth of an inch thick is left intact at the end of the vestibule. There is no other defensive precaution; no barricade, no heap of shavings. In order to come out, the insect has only to pierce an insignificant sheet of wood and then the bark.

The Nine-spotted Buprestis (*Ptosima novemmaculata*) behaves in the apricot-tree precisely as the Bronze Buprestis does in the poplar. Its larva bores the inside of the trunk with very low-ceilinged galleries, usually parallel with the axis; then, at a distance of an inch and a quarter or an inch and a half from the surface, it suddenly makes a sharp turn and proceeds in the direction of the bark. It tunnels straight ahead, taking the shortest road, instead of advancing by irregular windings as at first. Moreover, a sensitive intuition of coming events inspires its chisel to alter the plan of work. The perfect insect is a cylinder; the grub, wide in the thorax but slender elsewhere, is a strap, a ribbon. The first, with its unyielding cuirass, needs a cylindrical passage; the second needs a very low tunnel, with a roof that will give a purchase to the ambulatory nipples of the back. The larva therefore changes its manner of boring utterly: yesterday, the gallery, suited to a wandering life in the thickness of the wood, was a wide burrow with a very low ceiling, almost a slot; to-day the passage is cylindrical: a gimlet could not bore it more accurately. This sudden change in the system of road-making on behalf of the coming insect once more suggests for our meditation the eminent degree of foresight possessed by a bit of an intestine.

The cylindrical exit-way passes through the strata of wood along the shortest line, almost normally, after a slight bend which connects the vertical with the horizontal, a curve with a radius large enough to allow the stiff Buprestis to tack about without difficulty. It ends in a blind-alley, less than a twelfth of an inch from the surface of the wood. The eating away of the untouched sheet of wood and of the bark is all the labour that the grub leaves the insect to perform.

Having made these preparations, the larva withdraws, strengthening the wooden screen, however, with a layer of fine sawdust; it reaches the end of the round gallery, which is prolonged by the completely choked flat gallery; and here, scorning a special chamber or any upholstery, it goes to sleep for the nymphosis, with its head towards the exit.

I find numbers of specimens of a black Buprestis (*B. octoguttata*) in the old stumps of pine-trees left standing in the ground, hard outside but soft within, where the wood is as pliable as tinder. In this yielding substance, which has a resinous aroma, the larvæ spend their life. For the metamorphosis they leave the unctuous regions of the centre and penetrate the hard wood, where they hollow out oval recesses, slightly flattened, measuring from twenty-five to thirty millimetres[1] in length. The major axis of these cells is always vertical. They are continued by a wide exit-path, sometimes straight, sometimes slightly curved, according as the tree is to be quitted through the section above or through the side. The exit-channel is nearly always bored completely; the window by which the insect escapes opens directly upon the outside world. At most, in a few rare instances, the grub leaves the Buprestis the trouble of piercing a leaf of wood so thin as to be translucent. But, if easy paths are necessary to the insect, protective ramparts are no less needed for the safety of the nymphosis; and the larva plugs the liberating channel with a fine paste of masticated wood, very different from the ordinary sawdust. A layer of the same paste divides the bottom of the chamber from the low-ceilinged gallery, the work of the grub's active life. Lastly, the magnifying-glass reveals upon the walls of the cell a tapestry of woody fibres, very finely divided, standing erect and closely shorn, so as to make a sort of velvet pile. This quilted lining, of which the Cerambyx of the Oak showed us the first example, is, it seems to me, pretty often employed by the wood-eaters, Buprestes as well as Longicorns.

[1] .975 to 1.17 inch.—*Translator's Note*.

After these migrants, which travel from the centre of the tree to the surface, we will mention some others which from the surface plunge into the interior. A small Buprestis who ravages the cherry-trees, *Anthaxia nitidula*, passes his larval existence between the wood and the bark. When the time comes for changing its shape, the pigmy concerns itself, like the others, with future and present needs. To assist the perfect insect, the grub first gnaws the under side of the bark, leaving a thin screen of cuticle untouched, and then sinks in the wood a perpendicular well, blocked with unresisting sawdust. That is on behalf of the future: the frail Buprestis will be able to leave without hindrance. The bottom of the well, better wrought than the rest and ceiled with the aid of an adhesive fluid which holds the fine sawdust of the stopper in place, is a thing of the present; it is the nymphosis-chamber.

A second Buprestis, *Chrysobothrys chrysostigma*, likewise an exploiter of the cherry-tree, between the wood and the bark, although more vigorous, expends less labour on its preparations. Its chamber, with modestly varnished walls, is merely an expanded extension of the ordinary gallery. The grub, disinclined for persistent labour, does not bore the wood. It confines itself to hollowing a slanting dug-out in the bark, without touching the surface layer, through which the insect will have to gnaw its own way.

Thus each species displays special methods, tricks of the trade which cannot be explained merely by reference to its tools. As these minute details have consequences of some importance, I do not hesitate to multiply them: they all help to throw light upon the subject which we are investigating. Let us once more see what the Longicorns are able to tell us.

An inhabitant of old pine-stumps, *Criocephalus ferus* makes an exit-gallery which yawns widely on the outside world, opening either on the section of the stump or on the sides. The road is barricaded about two inches down with a long plug of coarse shavings. Next comes the nymph's cylindrical, compressed apartment, which is padded with woody fibres. It is continued underneath by the labyrinth of the larva, the burrow crammed full of digested wood. Note also the complete boring of the liberating passage, including the bark when there is any.

I find *Stromatium strepens* in ilex-logs which have been stripped of their bark. There is the same method of deliverance, the same passage curving gently towards the nearest outside point, the same barricade of shavings above the cell. Was the passage also carried through the bark? The stripped logs leave me ignorant as to this detail.

Clytus tropicus, a sapper of the cherry-tree, *C. arietis* and *C. arvicola*, sappers of the hawthorn, have a cylindrical exit-gallery, with a sharp turn to it. The gallery is masked on the outside by a remnant of bark or wood, hardly a millimetre thick,[2] and widens, not far from the surface, into a nymphosis-chamber, which is divided from the burrow by a mass of packed sawdust.

[2] .039 inch.—*Translator's Note.*

To continue the subject would entail an excess of monotonous repetition. The general law stands out very clearly from these few data: the wood-eating grubs of the Longicorns and Buprestes prepare the path of deliverance for the perfect insect, which will have merely in one case to pass a barricade of shavings or wormed wood, or in another to pierce a slight thickness of wood or bark. Thanks to a curious reversal of its usual attributes, youth is here the season of energy, of strong tools, of stubborn work; adult age is the season of leisure, of industrial ignorance, of idle diversions, without trade or profession. The infant has its paradise in the arms of its mother, its providence; here the infant, the grub, is the providence of the mother. With its patient tooth, which neither the perils of the outside world nor the difficult task of boring through hard wood are able to deter, it clears a way for her to the supreme delights of the sun. The youngster prepares an easy life for the adult.

Can these armour-wearers, so sturdy in appearance, be weaklings? I place nymphs of all the species that come to hand in glass tubes of the same diameter as the natal cell, lined with coarse paper, which will provide a good purchase for the boring. The obstacle to be pierced varies: a cork a centimetre thick;[3] a plug of poplar, very much softened by decay; a circular disk of sound wood. Most of my captives easily pierce the cork and the soft wood; these represent to them the barricade to be overthrown, the bark curtain to be perforated. A few, however, succumb before the front to be attacked; and all perish, after fruitless attempts, before the disk of hard wood. Thus perished the strongest of them all, the Great Capricorn, in my artificial oak-wood cells and even in my reed-stumps closed with their natural partitions.

[3] .39 inch.—*Translator's Note.*

They have not the strength, or rather the patient art; and the larva, more highly gifted, works for them. It gnaws with indomitable perseverance, an essential to success even for the strong; it digs with amazing foresight. It knows the future shape of the adult, whether round or oval, and bores the exit-passage accordingly, making it cylindrical in one case and elliptical in the other. It knows that the adult is very impatient to reach the light; and it leads her thither by the shortest way. In its wandering life in the heart of the tree, it loves low-roofed, winding tunnels, just big enough to pass through, or widening into stations when it strikes a vein with a better flavour; now, it makes a short, straight, roomy corridor, leading with a sharp bend to the outside world. It had plenty of time during its capricious wanderings; the adult has none to spare: his days are numbered; he must get out as quickly as he can. Hence the shortest road and as little encumbered by obstacles as is consistent with safety. The grub knows that the too sudden junction of the horizontal and the vertical part would stop the stiff, inflexible insect and bends it towards the outside with a gentle curve. This elbow changing the direction occurs whenever the larva ascends from the depths; it is very short when the nymphosis-chamber is next to the surface, but continues for some length when the chamber is well inside the trunk. In this case, the path traced by the grub has so regular a curve that you feel inclined to subject the work to geometrical measurement.

For want of sufficient data, I should have left this elbow in the shadow of a note of interrogation, had I had at my disposal only the emergence-galleries of the Longicorns and Buprestes, which are too short to lend themselves to trustworthy examination with the compasses. A lucky find provided me with the factors required. This was the trunk of a dead poplar, riddled, to a height of several yards, with an infinite number of round holes the diameter of a pencil. The precious pole, still standing, is uprooted with due respect, in view of my designs, and carried into my study, where it is sawn into longitudinal sections planed smooth.

The wood, while retaining its structure, has been greatly softened by the presence of the mycelium of a mushroom, the agaric of the poplar. The inside is decayed. The outer layers, to a depth of over four inches, are in good condition, save for the innumerable curved passages that cut through them. In a section involving the whole diameter of the trunk, the galleries of the late occupant produce a pleasing effect, of which a sheaf of corn gives us a pretty faithful image. Almost straight, parallel with one another and assembled in a bundle down the middle, they diverge at the top and spread into a cluster of wide curves, each of which ends in one of the holes on the surface. It is a sheaf of passages which has not the single head of a sheaf of corn, but shoots its innumerable sprouts hither and thither, at all heights.

I am enraptured by this magnificent specimen. The curves, of which I uncover a layer at every stroke of the plane, far exceed my requirements; they are strikingly regular; they afford the compasses the full space needed for accurate measurement.

Before calling in geometry, let us, if possible, name the creator of these beautiful curves. The inhabitants of the poplar have disappeared, perhaps long ago, as is proved by the mycelium of the agaric; the insect would not gnaw and bore its way through timber all permeated with the felt-like growth of the cryptogam. A few weaklings, however, have died without being able to escape. I find their remains swathed in mycelium. The agaric has preserved them from destruction by wrapping them in tight cerements. Under these mummy-bandages, I recognise a Saw-fly, *Sirex augur*, KLUG., in the state of the perfect insect. And—this is an important detail—all these adult remains, without a single exception, occupy spots which have no means of communication with the outside. I find them sometimes in a partly-constructed curved passage, beyond which the wood remains intact, sometimes at the end of the straight central gallery, choked with sawdust, which is not continued in front. These remains, with no thoroughfare before them, tell us plainly that the Sirex adopts for its exit methods not employed by the Buprestes and the Longicorns.

The larva does not prepare the path of deliverance; it is left for the perfect insect to open itself a passage through the wood. What I have before my eyes tells me more or less plainly the sequence of events. The larva, whose presence is proved by galleries blocked with packed sawdust, do not leave the centre of the trunk, a quieter retreat, less subject to the vicissitudes of the climate. Metamorphosis is effected at the junction of the straight gallery and the curved passage which is not yet made. When strength comes, the perfect insect tunnels ahead for a distance of more than four inches and opens up the exit-passage, which I find choked, not with compact sawdust, but with loose powdery rubbish. The dead insects which I strip of their mycelium-shrouds are weaklings whose strength deserted them mid-way. The rest of the passage is lacking because the labourer died on the road.

With this fact of the insect itself boring the exit passage, the problem assumes a more troublesome form. If the larva, rich in leisure and satisfied with its sojourn in the interior of the trunk, simplifies the coming emergence by shortening the road, what must not the adult do, who has so short a time to live and who is in so great a hurry to leave the hateful darkness? He above any other should be a judge of short cuts. To go from the murky heart of the tree to the sun-steeped bark, why does he not follow a straight line? It is the shortest way.

Yes, for the compasses, but not perhaps for the sapper. The length traversed is not the only factor of the work accomplished, of the total activity expended. We must take into account the resistance overcome, a resistance which varies according to the depth of the more or less hard strata and according to the method of attacking the woody fibres, which are either broken across or divided lengthwise. Under these conditions, whose value remains to be determined, can there be a curve involving a minimum of mechanical labour in cutting through the wood?

I was already trying to discover how the resistance may vary according to depth and direction; I was working out my differentials and my minimum integrals, when a very simple idea overturned my slippery scaffolding. The calculation of variations has nothing to do with the matter. The animal is not the moving body of the mathematicians, the particle of matter guided in its trajectory solely by the motive forces and the resistance of the medium traversed; it bears within itself conditions which control the others. The adult insect does not even enjoy the larva's privileges; it cannot bend freely in all directions. Under its harness it is almost a stiff cylinder. To simplify the explanation, we may liken the insect to a section of an inflexible straight line.

Let us return to the Sirex, reduced by abstraction to its axis. The metamorphosis is effected not far from the centre of the trunk. The insect lies lengthwise in the tree with its head up, very rarely with its head down. It must reach the outside as quickly as possible. The section of an inflexible straight line that represents it nibbles away a little wood in front of it and obtains a shallow cavity wide enough to allow of a very slight turn towards the outside. An infinitesimal advance is made; a second follows, the result of a similar cavity and a similar turn in the same direction. In short, each change of position is accompanied by the tiny deviation permitted by the slight excess of width of the hole; and this deviation invariably points the same way. Imagine a magnetic needle swung out of its position and tending to return to it while moving with a uniform speed through a resisting medium in which a sheath of a diameter slightly greater than the needle's opens bit by bit. The Sirex behaves more or less in the same fashion. His magnetic pole is the light outside. He makes for that direction by imperceptible deviations as his tooth digs.

The problem of the Sirex is now solved. The trajectory is composed of equal elements, with an invariable angle between them; it is the curve whose tangents, divided by infinitely small distances, retain the same inclination between each one and the next; the curve, in a word, with a constant angle of contingence. This characteristic betrays the circumference of the circle.

It remains to discover whether the facts confirm the logical argument. I take accurate tracings of a score of galleries, selecting those whose length best lends itself to the test of the compasses. Well, logic agrees with reality: over lengths which sometimes exceed four inches, the track of the compasses is identical with that of the insect. The most pronounced deviations do not exceed the small variations which we must reasonably expect in a problem of a physical nature, a problem incompatible with the absolute accuracy of abstract truths.

The Sirex' exit-gallery then is a wide arc of a circle whose lower extremity is connected with the corridor of the larva and whose upper extremity is prolonged in a straight line which ends at the surface with a perpendicular or slightly oblique incidence. The wide connecting arc enables the insect to tack about. When, starting from a position parallel with the axis of the tree, the Sirex has passed gradually to a transversal position, he completes his course in a straight line, which is the shortest road.

Does the trajectory imply the minimum of work? Yes, under the conditions of the insect's existence. If the larva had taken the precaution to place itself in a different direction when preparing for the nymphosis, to turn its head towards the nearest point of the bark instead of turning it lengthwise with the trunk, obviously the adult would escape more easily: he would merely have to gnaw straight in front of him in order to pass through the minimum thickness. But reasons of convenience whereof the grub is the sole judge, reasons dictated perhaps by weight, cause the vertical to precede the horizontal position. In order to pass from the former to the latter, the insect veers round by describing the arc of a circle. When this turn has been effected, the distance is completed in a straight line.

Let us consider the Sirex at his starting-point. His stiffness of necessity compels him to turn gradually. Here the insect can do nothing of its own initiative; everything is mechanically determined.

But, being free to pivot on its axis and to attack the wood on either side of the sheath, it has the option of attempting this reversal in a host of different ways, by a series of connected arcs, not in the same plane. Nothing prevents it from describing winding curves by revolving upon itself: spirals, loops constantly changing their direction, in fact, the complicated route of a creature that has lost its way. It might wander in a tortuous maze, making fresh attempts here, there and everywhere, groping for ever so long without succeeding.

But it does not grope and it succeeds very well. Its gallery is still contained within one plane, the first condition of the minimum of labour. Moreover, of the different vertical planes that can pass through the eccentric starting-point, one, the plane which passes through the axis of the tree, corresponds on the one side with the minimum of resistance to be overcome and on the other with the maximum. Nothing prevents the Sirex from tracing his path in any one of the multitude of planes on which the path would possess an intermediate value between the shortest and the longest. The insect refuses them all and constantly adopts the one which passes through the axis, choosing, of course, the side that entails the shortest path. In brief, the Sirex' gallery is contained in a plane pointing towards the axis of the tree and the starting-point; and of the two portions of this plane the channel passes through the less extensive. Under the conditions, therefore, imposed upon him by his stiffness the hermit of the poplar-tree releases himself with the minimum of mechanical labour.

The miner guides himself by the compass in the unknown depths underground, the sailor does the same in the unknown ocean solitudes. How does the wood-eating insect guide itself in the thickness of a tree-trunk? Has it a compass? One would almost say that it had, so successfully does it keep to the quickest road. Its goal is the light. To reach this goal, it suddenly chooses the economical plane trajectory, after spending its larval leisure in roaming tortuous passages full of irregular curves; it bends it in an arc which allows it to turn about; and, with its head held plumb with the adjacent surface, it goes straight ahead by the nearest way.

The most extraordinary obstacles are powerless to turn it aside from its plane and its curve, so imperative is its guiding force. It will gnaw metal, if need be, rather than turn its back upon the light, which it feels to be close at hand. The entomological records place this incredible fact beyond a doubt. At the time of the Crimean War, the Institut de France received some packets of cartridges in which the bullets had been perforated by *Sirex juvencus*; a little later, at the Grenoble Arsenal, *S. gigas* carved himself a similar exit. The larva was in the wood of the cartridge-boxes; and the adult insect, faithful to its direction of escape, had bored through the lead because the nearest daylight was behind that obstacle.

There is an exit-compass, that is incontestable, both for the larvæ preparing the passage of deliverance and for the adult insect, the Sirex obliged to make that passage for himself. What is it? Here the problem becomes surrounded with a darkness which is perhaps impenetrable; we are not well enough equipped with means of receiving impressions even to imagine the causes which guide the creature. There is, in certain events, another world of the senses in which our organs perceive nothing, a world which is closed to us. The eye of the camera sees the invisible and photographs the image of the ultra-violet rays; the tympanum of the microphone hears what to us is silence. A scientific toy, a chemical contrivance surpass us in sensibility. Would it be rash to attribute similar faculties to the delicate organization of the insect, even with regard to agencies unknown to our science, because they do not fall within the domain of our senses? To this question there is no positive reply; we have suspicions and nothing more. Let us at least dispel a few false notions that might occur to us.

Does the wood guide the insect, adult or larva, by its structure? Gnawed across the grain, it must produce a certain impression; gnawed lengthwise, it must produce a different impression. Is there not something here to guide the sapper? No, for in the stump of a tree left standing the emergence takes place, according to the proximity of the light, sometimes by way of the horizontal section, by means of a rectilinear path running along the grain, and sometimes by way of the side, by means of a curved road cutting across the grain.

Is the compass a chemical influence, or electrical, or calorific, or what not? No, for in an upright trunk the emergence is effected as often by the north face, which is always in the shade, as by the south face, which receives the sun all day long. The exit-door opens in the side which is nearest, without any other condition. Can it be the temperature? Not that either, for the shady side, though cooler, is utilized as often as the side facing the sun.

Can it be sound? Not so. The sound of what, in the silence of solitude? And are the noises of the outside world propagated through half an inch of wood in such a way as to make differences perceptible? Can it be weight? No again, for the trunk of the poplar shows us more than one Sirex travelling upside down, with his head towards the ground, without any change in the direction of the curved passages.

What then is the guide? I have no idea. It is not the first time that this obscure question has been put to me. When studying the emergence of the Three-pronged Osmia from the bramble-stems shifted from their natural position by my wiles, I recognized the uncertainty in which the evidence of physical science leaves us; and, in the impossibility of finding any other reply, I suggested a special sense, the sense of open space. Instructed by the Sirex, the Buprestes, the Longicorns, I am once again compelled to make the same suggestion. It is not that I care for the expression: the unknown cannot be named in any language. It means that the hermits in the dark know how to find the light by the shortest road; it is the confessions of an ignorance which no honest observer will blush to share. Now that the evolutionists' interpretations of instinct have been recognized as worthless, we all come to that stimulating maxim of Anaxagoras', which laconically sums up the result of my researches:

"[Greek: Nous pánta diekosmése]. Mind orders all things."

CHAPTER IX
THE DUNG-BEETLES OF THE PAMPAS

To travel the world, by land and sea, from pole to pole; to cross-question life, under every clime, in the infinite variety of its manifestations: that surely would be glorious luck for him that has eyes to see; and it formed the radiant dream of my young years, at the time when *Robinson Crusoe* was my delight. These rosy illusions, rich in voyages, were soon succeeded by dull, stay-at-home reality. The jungles of India, the virgin forests of Brazil, the towering crests of the Andes, beloved by the Condor, were reduced, as a field for exploration, to a patch of pebbles enclosed within four walls.

Heaven forfend that I should complain! The gathering of ideas does not necessarily imply distant expeditions. Jean-Jacques Rousseau[1] herborized with the bunch of chick-weed whereon he fed his Canary; Bernardin de Saint-Pierre[2] discovered a world on a strawberry-plant that grew by accident in a corner of his window; Xavier de Maistre,[3] using an arm-chair by way of post-chaise, made one of the most famous of journeys around his room.

[1] Jean-Jacques Rousseau (1712-1778), author of the *Confessions, La Nouvelle Héloïse,* etc.—*Translator's Note.*

[2] Jacques Henri Bernardin de Saint-Pierre (1737-1814), author of *Paul et Virginie, La Chaumière idienne* and *Etudes de la nature.*—*Translator's Note.*

[3] Xavier de Maistre (1763-1852), best known for his *Voyage autour de ma chambre* (1795).—*Translator's Note.*

This manner of seeing country is within my means, always excepting the post-chaise, which is too difficult to drive through the bushes. I go the circuit of my enclosure over and over again, a hundred times, by short stages; I stop here and I stop there; patiently, I put questions and, at long intervals, I receive some scrap of a reply.

The smallest insect village has become familiar to me: I know each fruit-branch where the Praying Mantis[4] perches; each bush where the pale Italian Cricket[5] strums amid the calmness of the summer nights; each downy plant scraped by the Anthidium, that maker of cotton bags; each cluster of lilac worked by the Megachile, the Leaf-cutter.

[4] Cf. *The Life of the Grasshopper:* chaps. vi. to ix.—*Translator's Note.*

[5] Cf. *idem:* chap. xvi.—*Translator's Note.*

If cruising among the nooks and corners of the garden do not suffice, a longer voyage shows ample profit. I double the cape of the neighbouring hedges and, at a few hundred yards, enter into relations with the Sacred Beetle,[6] the Capricorn, the Geotrupes,[7] the Copris,[8] the Decticus,[9] the Cricket,[10] the Green Grasshopper,[11] in short, with a host of tribes the telling of whose story would exhaust a lifetime. Certainly, I have enough and even too much to do with my near neighbours, without leaving home to rove in distant lands.

[6] Cf. *The Sacred Beetle and Others,* by J. Henri Fabre, translated by Alexander Teixeira de Mattos: chaps i. to vi.—*Translator's Note.*

[7] Cf. *idem:* chaps. xii. to xiv.—*Translator's Note.*

[8] Cf. *idem:* chaps. ix. and xvi.—*Translator's Note.*

[9] Cf. *The Life of the Grasshopper:* chaps. xi. to xiii.—*Translator's Note.*

[10] Cf. *idem:* chaps. xv. and xvi.—*Translator's Note.*

[11] Cf. *idem:* chap. xiv.—*Translator's Note.*

Besides, roaming the world, scattering one's attention over a host of subjects, is not observing. The travelling entomologist can stick numerous species, the joy of the collector and the nomenclator, into his boxes; but to gather circumstantial evidence is a very different matter. A Wandering Jew of science, he has no time to stop. Where a prolonged stay would be necessary to study this or that fact, he is hurried past the next stage. We must not expect the impossible of him under these conditions. Let him pin his specimens to cork tablets, let him steep them in jars of spirit, and leave to the sedentary the patient observations that require time.

This explains the extreme penury of history outside the dry descriptions of the nomenclator. Overwhelming us with its numbers, the exotic insect nearly always preserves the secret of its manners. Nevertheless, it were well to compare what happens under our eyes with that which happens elsewhere; it were excellent to see how, in the same guild of workers, the fundamental instinct varies with climatic conditions.

Then my longing to travel returns, vainer to-day than ever, unless one could find a seat on that carpet of which we read in the *Arabian Nights*, the famous carpet whereon one had but to sit to be carried whithersoever he pleased. O marvellous conveyance, far preferable to Xavier de Maistre's post-chaise! If I could only find just a little corner on it, with a return-ticket!

I do find it. I owe this unexpected good fortune to a Brother of the Christian Schools, to Brother Judulien, of the La Salle College at Buenos Aires. His modesty would be offended by the praises which his debtor owes him. Let us simply say that, acting on my instructions, his eyes take the place of mine. He seeks, finds, observes, sends me his notes and his discoveries. I observe, seek and find with him, by correspondence.

It is done; thanks to this first-rate collaborator, I have my seat on the magic carpet. Behold me in the pampas of the Argentine Republic, eager to draw a parallel between the industry of the Sérignan[12] Dung-beetles and that of their rivals in the western hemisphere.

[12] Sérignan, in Provence, where the author ended his days.—*Translator's Note.*

A glorious beginning! An accidental find procures me, to begin with, the Splendid Phanæus (*P. splendidulus*), who combines a coppery effulgence with the sparkling green of the emerald. One is quite astonished to see so rich a gem load its basket with ordure. It is the jewel on the dung-hill. The corselet of the male is grooved with a wide hollow and he sports a pair of sharp-edged pinions on his shoulders; on his forehead he plants a horn which vies with that of the Spanish Copris. While equally rich in metallic splendour, his mate has no fantastic embellishments, which are an exclusive prerogative of masculine dandyism among the Dung-beetles of La Plata as among our own.

Now what can the gorgeous foreigner do? Precisely what the Lunary Copris[13] does with us. Settling, like the other, under a flat cake of Cow-dung, the South American Beetle kneads egg-shaped loaves underground. Not a thing is forgotten: the round belly with the largest volume and the smallest surface; the hard rind which acts as a preservative against premature desiccation; the terminal nipple where the egg is lodged in a hatching-chamber; and, at the end of the nipple, the felt stopper which admits the air needed by the germ.

[13] Cf. *The Sacred Beetle and Others:* chap. xvi.—*Translator's Note.*

All these things I have seen here and I see over there, almost at the other end of the world. Life, ruled by inflexible logic, repeats itself in its works, for what is true in one latitude cannot be false in another. We go very far afield in search of a new spectacle to meditate upon; and we have an inexhaustible specimen before our eyes, between the walls of our enclosure.

Settled under the sumptuous dish dropped by the Ox, the Phanæus, one would think, ought to make the very best use of it and to stock her burrow with a number of ovoids, after the example of the Lunary Copris. She does nothing of the sort, preferring to roam from one find to the other and to take from each the wherewithal to model a single pellet, which is left to itself for the soil to incubate. She is not driven to practise economy even when she is working the produce of the Sheep far from the pastures of the Argentine.

The Glow-Worm and Other Beetles

Can this be because the jewel of the pampas dispenses with the father's collaboration? I dare not follow up the argument, for the Spanish Copris would give me the lie, by showing me the mother occupied alone in settling the family and nevertheless stocking her one pit with a number of pellets. Each has her share of customs the secret of which escapes us.

The two next, *Megathopa bicolor* and *M. intermedia,* have certain points of resemblance with the Sacred Beetle, for whose ebon hue they substitute a blue black. The first besides brightens his corselet with magnificent copper reflections. With their long legs, their forehead with its radiating denticulations and their flattened wing-cases, they are fairly successful smaller editions of the famous pill-roller.

They also share her talent. The work of both is once again a sort of pear, but constructed in a more ingenious fashion, with an almost conical neck and without any elegant curves. From the point of view of beauty, it falls short of the Sacred Beetle's work. Considering the tools, which have ample free play and are well adapted for clasping, I expected something better from the two modellers. No matter: the work of the Megathopæ conforms with the fundamental art of the other pill-rollers.

A fourth, *Bolbites onitoides,* compensates us for repetitions which, it is true, widen the scope of the problem but teach us nothing new. She is a handsome Beetle with a metallic costume, green or copper-red according as the light happens to fall. Her four-cornered shape and her long, toothed fore-legs make her resemble our Onites.[14]

[14] Cf. *The Sacred Beetle and Others:* chap. xvi.—*Translator's Note.*

In her, the Dung-beetles' guild reveals itself under a very unexpected aspect. We know insects that knead soft loaves; and here are some which, to keep their bread fresh, discover ceramics and become potters, working clay in which they pack the food of the larvæ. Before my housekeeper, before any of us, they knew how, with the aid of a round jar, to keep the provisions from drying during the summer heats. The work of the Bolbites is an ovoid, hardly differing in shape from that of the Copres; but this is where the ingenuity of the American insect shines forth. The inner mass, the usual dung-cake furnished by the Cow or the Sheep, is covered with a perfectly homogeneous and continuous coating of clay, which makes a piece of solid pottery impervious to evaporation.

The earthen pot is exactly filled by its contents, without the slightest interval along the line of junction. This detail tells us the worker's method. The jar is moulded on the provisions. After the food-pellet has been formed in the ordinary baker's fashion and the egg laid in its hatching-chamber, the Bolbites takes some armfuls of the clay near at hand, applies it to the foodstuff and presses it down. When the work is finished and smoothed to perfection with indefatigable patience, the tiny pot, built up piecemeal, looks as though made with the wheel and rivals our own earthenware in regularity.

The hatching-chamber, in which the egg lies, is, as usual, contrived in the nipple at the end of the pear. How will the germ and the young larva manage to breathe under that clay casing, which intercepts the access of the air?

Have no fears: the pot-maker knows quite well how matters stand. She takes good care not to close the top with the plastic earth which supplied her with the walls. At some distance from the tip of the nipple, the clay ceases to play its part and makes way for fibrous particles, for tiny scraps of undigested fodder, which, arranged one above the other with a certain order, form a sort of thatched roof over the egg. The inward and outward passage of the air is assured through this coarse screen.

One is set thinking in the presence of this layer of clay, which protects the fresh provisions, and this vent-hole stopped with a truss of straw, which admits the air freely, while defending the entrance. There is the eternal question, if we do not rise above the commonplace: how did the insect acquire so wise an art?

Not one fails in obeying those two laws, the safety of the egg and ready ventilation; not one, not even the next on my list, whose talent opens up a new horizon: I am now speaking of Lacordaire's Gromphas. Let not this repellant name of Gromphas (the old sow) give us a wrong notion of the insect. On the contrary, it is, like the last, an elegant Dung-beetle, dark-bronze, thickset, square-shaped like our Bison Onitis[15] and almost as large. It also practises the same industry, at least as regards the general effect of the work.

[15] Cf. *The Sacred Beetle and Others:* chap. xvi.—*Translator's Note.*

Its burrow branches into a small number of cylindrical cells, forming the homes of as many larvæ. For each of these the provisions consist of a parcel of Cow-dung, about an inch deep. The material is carefully packed and fills the bottom of the cavity, just as a soft paste would do when pressed down in a mould. Until now the work is similar to that of the Bison Onitis; but the resemblance goes no farther and is replaced by profound and curious differences, having no connection with what the Dung-beetles of our own parts show us.

As we know, our sausage-makers, Onites and Geotrupes alike, place the egg at the lower end of their cylinder, in a cell contrived in the very midst of the mass of foodstuffs. Their rival in the pampas adopts a diametrically opposite method: she places the egg above the victuals, at the upper end of the sausage. In order to feed, the grub does not have to work upwards; on the contrary, it works downwards.

More remarkable still: the egg does not lie immediately on top of the provisions; it is installed in a clay chamber with a wall about one-twelfth of an inch in thickness. This wall forms an hermetically-sealed lid, curves into a cup and then rises and bends over to make a vaulted ceiling.

The germ is thus enclosed in a mineral box, having no connection with the provision-store, which is kept strictly shut. The newborn grub must employ the first efforts of its teeth to break the seals, to cut through the clay floor and to make a trap-door which will take it to the underlying cake.

A rough beginning for the feeble mandible, even though the material to be bored through is a fine clay. Other grubs bite at once into a soft bread which surrounds them on every side; this one, on leaving the egg, has to make a breach in a wall before taking nourishment.

Of what use are these obstacles? I do not doubt that they have their purpose. If the grub is born at the bottom of a closed pot, if it has to chew through brick to reach the larder, I feel sure that certain conditions of its well-being demand this. But what conditions? To become acquainted with them would call for an examination on the spot; and all the data that I possess are a few nests, lifeless things very difficult to interrogate. However, it is possible to catch a glimpse of one or two points.

The Gromphas' burrow is shallow; those little cylinders, her loaves, are greatly exposed to drought. Over there, as here, the drying up of the victuals constitutes a mortal danger. To avert this peril, by far the most sensible course is to enclose the food in absolutely shut vessels.

Well, the receptacle is dug in very fine, homogeneous, water-tight earth, with not a bit of gravel, not an atom of sand in it. Together with the lid that forms the bottom of its round chamber, in which the egg is lodged, this cavity becomes an urn whose contents are safe from drought for a long time, even under a scorching sun. However late the hatching, the new-born grub, on finding the lid, will have under its teeth provisions as fresh as though they dated from that very day.

The clay food-pit, with its closely-fitting lid, is an excellent method, than which our agricultural experts have discovered no better way of preserving fodder; but it possesses one drawback: to reach the stack of food, the grub has first to open a passage through the floor of its chamber. Instead of the pap called for by its weakly stomach, it begins by finding a brick to chew.

The rude task would be avoided if the egg lay directly on top of the victuals, inside the case itself. Here our logic is at fault: it forgets an essential point, which the insect is careful not to disregard. The germ breathes. Its development requires air; and the perfectly-closed clay urn does not allow any air to enter. The grub has to be born outside the pot.

Agreed. But, in the matter of breathing, the egg is no better off for being shut up, on top of the provisions, in a clay casket quite as airtight as the jar itself. Examine the thing more closely, however, and you will receive a satisfactory reply. The walls of the hatching-chamber are carefully glazed inside. The mother has taken meticulous pains to give them a stucco-like finish. The vaulted ceiling alone is rugged, because the building-tool now works from the outside and is unable to reach the inner surface of the lid and smooth it. Moreover, in the centre of this curved and embossed ceiling, a small opening has been made. This is the air-hole, which allows of gaseous exchanges between the atmosphere inside the box and that outside.

If it were entirely free, this opening would be dangerous: some plunderer might take advantage of it to enter the casket. The mother foresees the risk. She blocks the breathing-hole with a plug made of the ravelled vegetable fibres of the Cow-dung, a stopper which is eminently permeable. It is an exact repetition of that which the various modellers have shown us at the top of their calabashes and pears. All of them are acquainted with the nice secret of the felt stopper as a means of ventilating the egg in a water-tight enclosure.

Your name is not an attractive one, my pretty Dung-beetle of the pampas, but your industrial methods are most remarkable. I know some among your fellow-countrymen, however, who surpass you in ingenuity. One of these is *Phanæus Milon,* a magnificent insect, blue-black all over.

The male's corselet juts forward. On the head is a short, broad, flattened horn, ending in a trident. The female replaces this ornament by simple folds. Both carry on the forehead two spikes which form a trusty digging-implement and also a scalpel for dissecting. The insect's squat, sturdy, four-cornered build resembles that of *Onitis Olivieri*, one of the rarities of the neighbourhood of Montpellier.

If similarity of shape implied purity of work, we ought unhesitatingly to attribute to *Phanæus Milon* short, thick puddings like those made by Olivier's Onitis.[16] Alas, structure is a bad guide where instinct is concerned! The square-chined, short-legged Dung-beetle excels in the art of manufacturing gourds. The Sacred Beetle herself supplies none that are more correctly shaped nor, above all, more capacious.

[16] I owe this detail on the work of Olivier's Onitis to a note and a sketch communicated by Professor Valéry-Mayer, of the Montpellier School of Agriculture.—*Author's Note*.

The thickset insect astonishes me with the elegance of its work, which is irreproachable in its geometry: the neck is shorter, but nevertheless combines grace with strength. The model seems derived from some Indian calabash, the more so as it has an open mouth and the belly is engraved with an elegant engine-turned pattern, produced by the insect's tarsi. One seems to see a pitcher protected by a wickerwork covering. The whole attains and even exceeds the size of a Hen's egg.

It is a very curious piece of work and of a rare perfection, especially when we consider the artist's clumsy and massive build. No, once again, the tool does not make the workman, among Dung-beetles any more than among ourselves. To guide the modeller there is something better than a set of tools: there is what I have called the bump, the genius of the animal.

Phanæus Milon scoffs at difficulties. He does much more than that: he laughs at our classifications. The word Dung-beetle implies a lover of dung. He sets no value on it, either for his own use or for that of his offspring. What he wants is the sanies of corpses. He is to be found under the carcasses of birds, Dogs or Cats, in the company of the undertakers-in-ordinary. The gourd which I will presently describe was lying in the earth under the remains of an Owl.

The Glow-Worm and Other Beetles

Let him who will explain this conjunction of the appetites of the Necrophorus[17] with the talents of the Sacred Beetle. As for me, baffled by tastes which no one would suspect from the mere appearance of the insect, I give it up.

[17] Or Burying-beetle. Cf. Chapters XI. and XII. of the present volume. — *Translator's Note.*

I know in my neighbourhood one Dung-beetle and one alone who also works among carrion. This is *Onthophagus ovatus*, LIN., a constant frequenter of dead Moles and Rabbits. But the dwarf undertaker does not on that account scorn stercoraceous fare: he feasts upon it like the other Onthophagi. Perhaps there is a twofold diet here: the bun for the adult; the highly-spiced, far-gone meat for the grub.

Similar facts are encountered elsewhere, with differing tastes. The Hunting Wasp takes her fill of honey drawn from the nectaries of the flowers, but feeds her little ones on game. Game first and then sugar, for the same stomach! How that digestive pouch must change during development! And yet no more than our own, which scorns in later life the food that delighted it when young.

Let us now examine the work of *Phanæus Milon* more thoroughly. The calabashes reached me in a state of complete desiccation. They are very nearly as hard as stone; their colour inclines to a pale chocolate. Neither inside nor out does the lens discover the slightest ligneous particle pointing to a vegetable residue. The strange Dung-beetle does not, therefore, use cakes of Cow-dung or anything like them; he handles products of another class, which at first are rather difficult to specify.

Held to the ear and shaken, the object rattles slightly, as would the shell of a dry fruit with a stone lying free inside it. Does it contain the grub, shrivelled by desiccation? Does it contain the dead insect? I thought so, but I was wrong. It contains something much more instructive than that.

The Glow-Worm and Other Beetles

I carefully rip up the gourd with the point of a knife. Within a homogenous wall, whose thickness is over three-quarters of an inch in the largest of my three specimens, is encased a spherical kernel, which fills the cavity exactly, but without sticking to the wall at any part. The small amount of free play allowed to this kernel accounts for the rattling which I heard when I shook the thing.

In the colour and general appearance of the whole, the kernel does not differ from the wrapper. But break it open and minutely examine the pieces. We now recognize tiny fragments of bone, flocks of down, threads of wool, scraps of flesh, the whole mixed in an earthy paste resembling chocolate.

This paste, when placed on hot charcoal, sifted under the lens and deprived of its particles of dead bodies, becomes much darker, is covered with shiny bubbles and sends forth puffs of that acrid smoke by which we so readily recognize burnt animal matter. The whole mass of the kernel, therefore, is strongly impregnated with sanies.

Treated in the same manner, the wrapper also turns black, but not to the same extent; it hardly smokes; it does not become covered with jet-black bubbles; lastly, it would not anywhere contain bits of carcase similar to those in the central kernel. In both cases, the residue after calcination is a fine, reddish clay.

This brief analysis tells us all about the table of *Phanæus Milon*. The fare served to the grub is a sort of meat-pie. The sausage-meat consists of a mince of all that the two scalpels of the forehead and the toothed knives of the fore-legs have been able to remove from the corpse: hair and down, small crushed bones, strips of flesh and skin. Now hard as brick, the thickening of this mincemeat was originally a paste of fine clay steeped in the liquor of corruption. Lastly, the light crust of our meat-pies is here represented by a covering of the same clay, less rich in extract of meat than the other.

The pastry-cook gives his work an elegant shape; he decorates it with rosettes, with twists, with scrolls. *Phanæus Milon* is no stranger

The Glow-Worm and Other Beetles

to these culinary æsthetics. She turns the crust of her meat-pie into a splendid gourd, with a finger-print ornamentation.

The outer covering, an unprofitable crust, insufficiently steeped in savoury juices, is not, we can easily guess, intended for consumption. It is possible that, somewhat later, when the stomach becomes robust and is not repelled by coarse fare, the grub scrapes a little from the sides of its pasty walls; but, until the adult insect emerges, the calabash as a whole remains intact, having acted at first as a safeguard of the freshness of the force-meat and all the while as a protecting casket for the recluse.

Above the cold pastry, right at the base of the neck of the gourd, is contrived a round cell with a clay wall continuing the general wall. A fairly thick floor, made of the same material, separates it from the store-room. This is the hatching-chamber. Here is laid the egg, which I find in its place but dried up; here is hatched the grub, which, to reach the ball of food, must first open a trap-door through the partition that separates the two stories.

We have here, in short, the edifice of the Gromphas, in a different style of architecture. The grub is born in a casket surmounting the stack of food but not communicating with it. The budding larva must therefore, at the opportune moment, itself pierce the covering of the pot of preserves. As a matter of fact, later, when the grub is on the sausage-meat, we find the floor perforated with a hole just large enough for it to pass through.

Wrapped all round in a thick casing of pottery, the meat keeps fresh as long as is required by the duration of the hatching-process, a detail which I have not ascertained; in its cell, which is also of clay, the egg lies safe. Capital; so far, all is well. *Phanæus Milon* is thoroughly acquainted with the secrets of fortification and the danger of victuals evaporating too soon. There remain the germ's respiratory requirements.

To satisfy these, the insect has been equally well-inspired. The neck of the calabash is pierced, in the direction of its axis, with a tiny channel which would admit at most the slenderest of straws. Inside, this conduit opens at the top of the dome of the hatching-chamber; outside, at the tip of the nipple, it spreads into a wide mouth. This is the ventilating-shaft, protected against intruders by its extreme narrowness and by grains of dust which obstruct it a little without stopping it up. I said it was simply marvellous. Was I wrong? If a construction of this sort is a fortuitous result, we must admit that blind chance is gifted with extraordinary powers of foresight.

How does the clumsy insect manage to accomplish so delicate and complex a piece of building? Exploring the pampas as I do through the eyes of an intermediary, my only guide in this question is the structure of the work, a structure whence we can deduct the workman's method without going far astray. I therefore imagine the building to proceed in this manner: a small carcase is found, the oozing of which has softened the underlying loam. The insect collects more or less of this loam, according to the richness of the vein. There are no precise limits here. If the plastic material be plentiful, the collector is lavish with it and the provision-box becomes all the more solid. Then enormous calabashes are obtained, exceeding a Hen's egg in volume and formed of an outer wall three-quarters of an inch thick. But a mass of this description is beyond the strength of the modeller, is badly handled and betrays, in its shape, the awkwardness attendant on an over-difficult task. If the material be rare, the insect confines its harvesting to what is strictly necessary; and then, freer in its movements, it obtains a magnificently regular gourd.

The loam is probably first kneaded into a ball and then scooped out into a large and very thick cup by the pressure of the fore-legs and the work of the forehead. Even thus do the Copris and the Sacred Beetle act when preparing, on the top of their round pill, the bowl in which the egg will be laid before the final manipulation of the ovoid or pear.

In this first business, the Phanæus is simply a potter. So long as it be plastic, any clay serves her turn, however meagrely saturated with the juices running from the carcase.

She now becomes a pork-butcher. With her toothed knife, she carves, she saws some tiny shreds from the rotten animal; she tears off, cuts away what she deems best suited to the grub's entertainment. She collects all these fragments and mixes them with choice loam in the spots where the sanies abounds. The whole, cunningly kneaded and softened, becomes a ball made on the spot, without any rolling-process, in the same way as the sphere of the other pill-manufacturers. Let us add that this ball, a ration calculated by the needs of the grub, is very nearly constant in size, whatever the dimensions of the final calabash.

The sausage-meat is now ready. It is set in place in the wide-open clay bowl. Loosely packed, without compression, the food will remain free, will not stick to its wrapper.

Next, the potter's work is renewed. The insect presses the thick lips of the clay cup, rolls them out and applies them to the prepared force-meat, which is eventually contained by a thin partition at the top end and by a thick layer every elsewhere. A wide circular pad is left on the top partition, which is thin in view of the weakness of the grub that is to perforate it later, when making for the provisions. Manipulated in its turn, this pad is converted into a hemispherical hollow, in which the egg is forthwith laid.

The work is completed by rolling out and joining the edges of the little crater, which closes and becomes the hatching-chamber. Here, especially, a delicate dexterity becomes essential. At the same time that the nipple of the calabash is being shaped, the insect, when packing the material, must leave the little channel which is to form the ventilating-shaft, following the line of the axis. This narrow conduit, which an ill-calculated pressure might stop up beyond hope of remedy, seems to me extremely difficult to obtain. The most skilful of our potters could not manage it without the aid of a needle, which he would afterwards withdraw. The insect, a sort of jointed automaton, makes its channel through the massive nipple of the gourd without so much as a thought. If it did give it a thought, it would not succeed.

The calabash is made: there remains the decoration. This is the work of patient after-touches which perfect the curves and leave on the soft loam a series of stippled impressions similar to those which the potter of prehistoric days distributed over his big-bellied jars with the ball of his thumb.

That finishes the work. The insect will begin all over again under a fresh carcase; for each burrow has one calabash and no more, even as with the Sacred Beetle and her pears.

Here is another of these artists of the pampas. All black and as big as the largest of our Onthophagi,[18] whom she greatly resembles in general build, *Canthon bispinus* is likewise an exploiter of dead bodies, if not always on her own behalf, at least on that of her offspring.

[18] Cf. *The Sacred Beetle and Others:* chaps. xi. xvii., and xviii.—*Translator's Note.*

She introduces very original innovations into the pill-maker's art. Her work, strewn like the aforementioned with finger-prints, is the pilgrim's gourd, the double-bellied gourd. Of the two stories, which are joined together by a fairly plainly-marked groove, the upper is the smaller and contains the egg in an incubating-chamber; the lower and bulkier is the food-stack.

Imagine the Sisyphus' little pear with its hatching-chamber swollen into a globule a trifle smaller than the sphere at the other end; suppose the two protuberances to be divided by a sort of wide open groove like that of a pulley; and we shall have something very like the Canthon's work in shape and size.

When placed on burning charcoal, this double-bellied gourd turns black, becomes covered with shiny warts that look like jet beads, emits a smell like that of grilled meat and leaves a residue of red clay. It is therefore formed of clay and sanies. Moreover, the paste is sprinkled with little scraps of dead flesh. At the smaller end is the egg, in a chamber with a very porous roof, to allow the air to enter.

The Glow-Worm and Other Beetles

The little undertaker has something better to show than her double sausage. Like the Bison Onitis, the Sisyphus and the Lunary Copris, she enjoys the collaboration of the father. Each burrow contains several cradles, with the father and mother invariably present. What are the two inseparables doing? They are watching their brood and, by dint of assiduous repairs, keeping the little sausages, which are in constant danger of cracking or drying up, in good condition.

The magic carpet which has allowed me to take this trip to the pampas supplies me with nothing else worth noting. Besides, the New World is poor in pill-rollers and cannot compare with Senegambia and the regions of the Upper Nile, that paradise of Copres and Sacred Beetles. Nevertheless we owe it one precious detail: the group which is commonly known by the name of Dung-beetles is divided into two corporations, one of which exploits dung, the other corpses.

With very few exceptions, the latter has no representatives in our climes. I have mentioned the little Oval Onthophagus as a lover of carrion corruption; and my memory does not recall any other example of the kind. We have to go to the other world to find such tastes.

Can it be that there was a schism among the primitive scavengers and that these, at first addicted to the same industry, afterwards divided the hygienic task, some burying the ordure of the intestines, the others the ordure of death? Can the comparative frequency of this or the other provender have brought about the formation of two trade-guilds?

That is not admissible. Life is inseparable from death; wherever a corpse is, there also, scattered at random, are the digestive residues of the live animal; and the pill-roller is not fastidious as to the origin of this waste matter. Dearth therefore plays no part in the schism, if the true dung-worker has actually turned himself into an undertaker, or if the undertaker has turned himself into a true dung-worker. At no time have materials for the work been lacking in either case.

Nothing, not the scarcity of provisions, nor the climate, nor the reversed seasons, would explain this strange divergence. We must perforce regard it as a matter of original specialities, of tastes not acquired but prescribed from the beginning. And what prescribed them was anything but the structure.

I would defy the greatest expert to tell me, simply from the insect's appearance and without learning the facts by experiment, the manner of industry to which *Phanæus Milon*, for instance, devotes himself. Remembering the Onites, who are very similar in shape and who manipulate stercoral matter, he would look upon the foreigner as another manipulator of dung. He would be mistaken: the analysis of the meat-pie has told us so.

The shape does not make the real Dung-beetle. I have in my collection a magnificent insect from Cayenne, known to the nomenclators as *Phanæus festivus*, a brilliant Beetle in festive attire, charming, beautiful, glorious to behold. How well he deserves his name! His colouring is a metallic red, which flashes with the fire of rubies; and he sets off this splendid jewellery by studding his corselet with great spots of glowing black.

What trade do you follow under your torrid sun, O gleaming carbuncle? Have you the bucolic tastes of your rival in finery, the Splendid Phanæus? Can you be a knacker, a worker in putrid sausage-meat, like *Phanæus Milon*? Vainly do I consider you and marvel at you: your equipment tells me nothing. No one who has not seen you at work is capable of naming your profession. I leave the matter to the conscientious masters, to the experts who are able to say: I do not know. They are scarce, in our days; but after all there are some, less eager than others in the unscrupulous struggle which creates upstarts.

This excursion to the pampas leans to one conclusion of some importance. We find in another hemisphere, with reversed seasons, a different climate and dissimilar biological conditions, a series of true dung-workers whose habits and industry repeat, in their essential facts, the habits and industry of our own. Prolonged investigations,

The Glow-Worm and Other Beetles

made at first hand and not, like mine, at second hand, would add greatly to the list of similar workers.

And it is not only in the grassy plains of La Plata that the modellers of dung proceed according to the principles usual over here; we may say, without fear of being mistaken, that the magnificent Copres of Ethiopia and the big Sacred Beetles of Senegambia work exactly like our own.

The same similarity of industry exists in other entomological series, however distant their country. My books give details of a Pelopæus[19] in Sumatra, who is an ardent Spider-huntress like our own, who builds mud cells inside houses and who, like her, is fond of the loose hangings of the window-curtains for the shifting foundation of her nests. They tell me of a Scolia[20] in Madagascar who serves each of her grubs with a fat rasher, an Oryctes-larva,[21] even as our own Scoliæ feed their family on prey of similar organization, with a highly concentrated nervous system, such as the larvæ of Cetoniæ, Anoxiæ and even Oryctes. They tell me that in Texas a Pepsis, a huntress of big game akin to the Calicurgi, gives chase to a formidable Tarantula and vies in daring with our Ringed Calicurgus,[22] who stabs the Black-bellied Lycosa.[23] They tell me that the Sphex-wasps of the Sahara, a rival of our own White-banded Sphex,[24] operate on Locusts. But we must limit these quotations, which could easily be multiplied.

[19] Cf. *The Mason-wasps,* by J. Henri Fabre, translated by Alexander Teixeira de Mattos: chaps. iii. to vi.—*Translator's Note.*

[20] The chapters on the Scoliæ will appear in *More Hunting Wasps.* Meanwhile, cf. *The Life and Love of the Insect,* by J. Henri Fabre, translated by Alexander Teixeira de Mattos: chap. xi.—*Translator's Note.*

[21] The larva of the Rhinoceros Beetle.—*Translator's Note.*

[22] For the Pompilus, or Ringed Calicurgus, cf. *The Life and Love of the Insect:* chap. xii.—*Translator's Note.*

[23] For the Narbonne Lycosa, or Black-bellied Tarantula, cf. *The Life of the Spider:* chaps. i. and iii. to vii.—*Translator's Note.*

[24] Cf. *The Life of the Fly:* chap. i.—*Translator's Note.*

For producing variations of animal species to suit our theorists there is nothing so convenient as the influence of environment. It is a vague, elastic phrase, which does not compromise us by compelling us to be too precise and it supplies an apparent explanation of the inexplicable. But is this influence so powerful as they say?

I grant you that to some small extent it modifies the shape, the fur or feather, the colouring, the outward accessories. To go farther would be to fly in the face of facts. If the surroundings become too exacting, the animal protests against the violence endured and succumbs rather than change. If they go to work gently, the creature subjected to them adapts itself as best it can, but invincibly refuses to cease to be what it is. It must live in the form of the mould whence it issued, or it must die: there is no other alternative.

Instinct, one of the higher characteristics, is no less rebellious to the injunctions of environment than are the organs, which serve its activity. Innumerable guilds divide the work of the entomological world; and each member of one of these corporations is subject to rules which not climate, nor latitude, nor the most serious disturbances of diet are able to alter.

Look at the Dung-beetles of the pampas. At the other end of the world, in their vast flooded pastures, so different from our scanty greenswards, they follow, without notable variations, the same methods as their colleagues in Provence. A profound change of surroundings in no way effects the fundamental industry of the group.

Nor do the provisions available affect it. The staple food to-day is matter of bovine origin. But the Ox is a newcomer in the land, an importation of the Spanish conquest. What did the Megathopæ, the Bolbites, the Splendid Phanæus eat and knead, before the arrival of the present purveyor? The Llama, that denizen of the uplands, was not able to feed the Dung-beetles confined to the plains. In days of old, the foster-father was perhaps the monstrous Megatherium, a dung-factory of incomparable prodigality.

And from the produce of the colossal beast, whereof naught remains but a few rare skeletons, the modellers passed to the produce of the Sheep and the Ox, without altering their ovoids or their gourds, even as our Sacred Beetle, without ceasing to be faithful to her pear, accepts the Cow's flat cake in the absence of the favourite morsel, the Sheep's bannock.

In the south as in the north, at the antipodes as here, every Copris fashions ovoids with the egg at the smaller end; every Sacred Beetle models pears or gourds with a hatching-chamber in the neck; but the materials employed vary greatly according to the season and locality and can be furnished by the Megatherium, the Ox, the Horse, the Sheep or by man and several others.

We must not allow this diversity to lead us to believe in changes of instinct: that would be to strain at a Gnat and swallow a Camel. The industry of the Megachiles, for instance, consists of manufacturing wallets with bits of leaves; that of the Cotton-bees of making bags of wadding with the flock gathered from certain plants. Whether the pieces be cut from the leaves of this shrub or that, or at need from the petals of some flower; whether the cotton-wool be collected here or there, as chance may direct the encounter, the industry undergoes no essential changes.

In the same manner, nothing changes in the art of the Dung-beetle, victualling himself with materials in this mine or that. Here in truth we have immutable instinct, here we behold the rock which our theorists are unable to shake.

And why should it change, this instinct, so logical in its workings? Where could it find, even with chance assisting, a better plan? In spite of an equipment which varies in the different genera, it suggests to every modelling Dung-beetle the spherical shape, a fundamental structure which is hardly affected when the egg is placed in position.

From the outset, without the use of compasses, without any mechanical rolling, without shifting the thing on its base, one and all obtain the ball, the delicately executed compact body supremely favourable to the grub's well-being. To the shapeless lump, demanding no pains, they all prefer the sphere, lovingly fashioned and calling for much manipulation, the globe which is the preeminent form and best-adapted for the preservation of energy, in the case of a sun and of a Dung-beetle's cradle alike.

When Macleay[25] gave the Sacred Beetle the name of Heliocantharus, the Black-beetle of the Sun, what had he in mind? The radiating denticulations of the forehead, the insect's gambols in the bright sunlight? Was he not thinking rather of the symbol of Egypt, the Scarab who, on the pediment of the temples, lifts towards the sky, by way of a pill, a vermilion sphere, the image of the sun?

[25] William Sharp Macleay (1792-1865), author of *Horæ Entomologicæ; or, Essays on Annulose Animals* (1819-1821), on which I quote the *Dictionary of National Biography*:
"He propounded the circular or quinary system, a forcedly artificial attempt at a natural system of classification, which soon became a byword among naturalists." — *Translator's Note*.

The comparison between the mighty bodies of the universe and the insect's humble pellet was not distasteful to the thinkers on the banks of the Nile. For them supreme splendour found its effigy in extreme abjection. Were they very wrong?

No, for the pill-roller's work propounds a grave problem to whoso is capable of reflection. It compels us to accept this alternative: either to credit the Dung-beetle's flat head with the signal honour of having of itself solved the geometrical problem of preserved food, or else to fall back upon a harmony ruling all things under the eye of an Intelligence Which, knowing everything, has provided for everything.

CHAPTER X
INSECT COLOURING

Phanæus splendidulus, the glittering, the resplendent: this is the epithet selected by the official nomenclators to describe the handsomest Dung-beetle of the pampas. The name is not at all exaggerated. Combining the fire of gems with metallic lustre, the insect, according to the incidence of the light, emits the green reflections of the emerald or the gleam of ruddy copper. The muck-raker would do honour to the jeweller's show-cases.

For the rest, our own Dung-beetles, though usually modest in their attire, also have a leaning toward luxurious ornament. One Onthophagus decorates his corselet with Florentine bronze; another wears garnets on his wing-cases. Black above, the Mimic Geotrupes is the colour of copper pyrites below; also black in all parts exposed to the light of day, the Stercoraceous Geotrupes displays a ventral surface of a glorious amethyst violet.

Many other series, of greatly varied habits, Carabi,[1] Cetoniæ, Buprestes, Chrysomelæ,[2] rival and even surpass the magnificent Dung-beetles in the matter of jewellery. At times we encounter splendours which the imagination of a lapidary would not venture to depict. The Azure Hoplia,[3] the inmate of the osier-beds and elders by the banks of the mountain streams, is a wonderful blue, tenderer and softer to the eye than the azure of the heavens. You could not find an ornament to match it save on the throats of certain Humming-birds and the wings of a few Butterflies in equatorial climes.

[1] Cf. Chapter XIV. of the present volume.—*Translator's Note.*

[2] Golden Apple-beetles.—*Translator's Note.*

[3] A genus of Cockchafer. Cf. *The Life of the Fly:* chap. vii.—*Translator's Note.*

To adorn itself like this, in what Golconda does the insect gather its gems? In what diggings does it find its gold nuggets? What a pretty problem is that of a Buprestis' wing-case! Here the chemistry of colours ought to reap a delightful harvest; but the difficulties are great, it seems, so much so that science cannot yet tell us the why and the wherefore of the humblest costume. The answer will come in a remote future, if indeed it ever comes completely, for life's laboratory may well contain secrets denied to our retorts. For the moment, I shall perhaps be contributing a grain of sand to the future palace if I describe the little that I have seen.

My basic observation dates a long way back. I was at that time busy with the Hunting Wasps, following their larval development from the egg to the cocoon. Let us take an instance from my notes, which cover nearly all the game-hunters of my district. I will choose the larva of the Yellow-winged Sphex,[4] which, with its convenient size, will furnish an easy object-lesson.

[4] Cf. *The Hunting Wasps:* chap. iv.—*Translator's Note.*

Under the transparent skin of the larva, which has been recently hatched and is consuming its first Cricket, we soon perceive some fine white spots, which rapidly increase in size and number and eventually cover the whole body, except the first two or three segments. On dissecting the grub, we find that these spots have to do with the adipose layer, of which they form a considerable part, for, far from being scattered only on the surface, they run through its whole thickness and are present in such numbers that the forceps cannot seize the least fragment of this tissue without picking up a few of them.

Though perfectly visible without the help of a lens, these puzzling spots call for the microscope if we wish to study them in detail. We then find that the adipose tissue is made up of two kinds of vesicles: some, bright yellow and transparent, are filled with oily drops; the rest, opaque and starch-white, are distended with a very fine powder, which spreads in a cloudy trail when the vesicle containing it is broken on the object-slide. Intermingled without apparent order, the two kinds of bags are of the same shape and the same size. The first go to make up the nutritive reserves, the fatty tissue properly so-called; the second form the white dots which we will study for a moment.

An inspection under the microscope tells us that the contents of the white cells are composed of very fine, opaque grains, insoluble in water and of greater density. The use of chemical reagents on the object-slide proves that nitric acid dissolves these grains, with effervescence and without leaving the least residue, even when they are still enclosed in their vesicles. On the other hand, the true fatty cells suffer in no way when attacked by this acid; they merely turn a little yellower.

Let us take advantage of this property to operate on a larger scale. The adipose tissue taken from a number of larvæ is treated with nitric acid. The effervescence is as lively as if the reaction were taking effect on a bit of chalk. When it has subsided, some yellow clots are floating on the surface. These are easily separated. They come from the fatty substance and the cellular membranes. There remains a clear liquid containing the white granules in solution.

The riddle of these granules was being presented to me for the first time; my predecessors had provided no physiological or anatomical data to guide me; great therefore was my joy when, after a little fumbling, I succeeded in hitting upon their characteristic feature.

The solution is evaporated in a small porcelain capsule, placed on the hot embers. On the residue I pour a few drops of ammonia, or else simply water. A glorious crimson colour at once makes its appearance. The problem is solved: the colouring-matter which has just formed is murexide; and consequently the powdery substance which filled the cells was none other than uric acid, or more precisely ammonium urate.

A physiological fact of this importance can hardly stand alone. Indeed, since this basic experiment I have discovered grains of uric acid in the adipose tissue of the larvæ of all the Hunting Wasps of our parts, as well as in the Bees at the moment of the nymphosis. I have observed them in many other insects, either in the larval or in the perfect state; but in this respect there is none to equal the grub of the game-hunting Wasp, which is all speckled with white. I think I see the reason.

Let us consider two larvæ which eat live prey: that of the Sphex and that of the Hydrophilus.[5] Uric acid, the inevitable product of the vital transformations, or at all events one of its analogues, must be formed in both. But the Hydrophilus' larva shows no accumulation of it in its adipose layer, whereas the Sphex' is full of it.

[5] The Great Water-beetle.—*Translator's Note.*

In the latter the duct through which the solid excretions pass is not yet in working order; the digestive apparatus, tied at the lower end, is not discharging an atom. The urinary products, being unable, for want of an open outlet, to flow away as formed, accumulate in the adipose tissue, which thus serves as a common store-house for the residues of the present and the plastic material of the future organic processes. Here something occurs analogous to what we see in the higher animals after the removal of the kidneys; the urea at first contained in the blood, in imperceptible quantities accumulates and becomes manifest when the means by which it is eliminated disappear.

In the larva of the Hydrophilus, on the other hand, the excretions enjoy a free outlet from the beginning; and the urinary products escape as and when formed and are no longer deposited in the adipose tissue. But during the intense labour of the metamorphosis, any excretion becomes impossible; the uric acid must and does collect in the adipose substance of the different larvæ.

It would be out of place, despite its importance, to pursue the problem of the uric residues any further. Our subject is coloration. Let us return to it with the evidence supplied by the Sphex. Her almost transparent larva has the neutral tint of fluid white of egg. Under its fine translucid skin there is nothing coloured, save the long digestive pouch, which is swollen a deep purple by the pulp of the consumed Crickets. But against this indefinite, vitreous background the opaque white uric cells stand out distinctly in their myriads; and the effect of this stippling is a sketchy but by no means inelegant costume. It is skimpy in the extreme, but at any rate it is something.

With the urinary broth of which its intestine is unable to get rid, the larva has discovered a means of making itself look a little smart. The Anthidia have shown us how, in their cotton-wool wallets, they manufacture a sort of jewellery with their ordure. The robe studded with grains of alabaster is a no less ingenious invention.

To beautify themselves cheaply by using up their own refuse is a very common method even among insects endowed with all that is wanted for evacuating waste matter. While the larvæ of the Hunting Wasps, unable to do better, stipple themselves with uric acid, there are plenty of industrious creatures that are able to make themselves a superb dress by preserving their excretions in spite of their own open sewers. With a view to self-embellishment, they collect and treasure up the dross which others hasten to expel. They turn filth into finery.

One of these is the White-faced Decticus (*D. albifrons*, FAB.), the biggest sabre-bearer of the Provençal fauna. A magnificent insect is this Grasshopper, with a broad ivory face, a full, creamy-white belly and long wings flecked with brown. In July, the season for the wedding-dress, let us dissect him under water.

The adipose tissue, which is abundant and yellowish white, consists of a lace of wide, irregular, criss-cross meshes. It is a tubular network swollen with a powdery matter which condenses into minute chalk-white spots, standing out very plainly against a transparent background. When crushed in a drop of water, a fragment of this fabric yields a milky cloud in which the microscope shows an infinite number of opaque floating atoms, without revealing the smallest blob of oil, the sign of fatty matter.

The Glow-Worm and Other Beetles

Here again we have ammonium urate. Treated with nitric acid, the adipose tissue of the Decticus produces an effervescence similar to that of chalk and yields enough murexide to redden a tumblerful of water. A strange adipose body, this bundle of lace crammed with uric acid without a trace of fatty matter! What would the Decticus do with nutritive reserves, seeing that he is near his end, now that the nuptial season has arrived? Delivered from the necessity of saving for the future, he has only to spend in gaiety the few days left to him; he has only to adorn himself for the supreme festival.

He therefore converts into a paint-factory what at first was a warehouse for storing up foodstuffs; and with his chalk-like uric pulp he lavishly daubs his belly, which turns a creamy white, and smears it on his forehead, his face, his cheeks, until they assume the appearance of old ivory. All those parts, in fact, which lie immediately under the translucid skin are covered with a layer of pigment which can be turned into murexide and is identical in nature with the white powder of the adipose lace.

Biological chemistry can hardly offer a simpler and more striking experiment than this analysis of the Decticus' finery. To those who have not this curious Grasshopper handy, I recommend the Ephippiger of the Vines, who is much more widely distributed. His ventral surface, which also is of a creamy white, likewise owes its colour to a plastering of uric acid. In the Grasshopper family many other species of smaller size and requiring more delicate handling would give us the same results in varying degrees.

White, slightly tinged with yellow, is all that the urinary palette of the Locustidæ shows us. A caterpillar, the Spurge Hawk-moth's, will take us a little farther. Dappled red, black, white and yellow, its livery is the most remarkable in our part of the country. Réaumur in fact calls it *la Belle*. The flattering title is well-deserved. On the black background of the larva, vermillion-red, chrome-yellow and chalk-white figure side by side in circles, spots, freckles and stripes, as clearly marked as the glaring patches of a harlequin's dress.

Let us dissect the caterpillar and apply the lens to its mosaic. On the inner surface of the skin, except in the portions coloured black, we observe a pigmentary layer, a coating here red, there yellow or white. We will cut a strip from this coat of many colours, after depriving it of its muscular fibres, and subject it to the action of nitric acid. The pigment, no matter what its hue, dissolves with effervescence and afterwards yields murexide. Here again, then, it is to uric acid, present, however, in small quantities in the adipose tissue, that the caterpillar's rich livery is due.

The black parts are an exception. Unassailable by nitric acid, they retain their sombre tint after treatment as before, whereas the portions stripped of their pigment by the reagent become almost as transparent as glass. The skin of the handsome caterpillar thus has two sorts of coloured patches.

Those of an intense black may be likened to dyers' products: they are completely impregnated with the colouring matter, which is part and parcel of the molecular constitution and cannot be isolated by the nitric solvent. The others, red, yellow or white, are actually painted: on a translucid sheet is a wash of urinary pigment, which is discharged by the minute ducts issuing from the adipose layer. When the action of the nitric acid has ceased, the transparent circles of the latter stand out against the black background of the former.

Yet one more example taken from a different order. As regards elegance of costume, the Banded Epeira[6] is the most highly favoured of our Spiders. On the upper surface of her corpulent belly alternate, in transversal bands, bright black, a vivid yellow like that of yolk of egg and a dazzling white like that of snow. The black and yellow also show underneath, but arranged differently. The yellow, in particular, forms two longitudinal ribbons, ending in orange-red on either side of the spinnerets. A pale purple is faintly diffused over the sides.

[6] Cf. *The Life of the Spider:* chaps. ii., vii., xi. and xiii.—*Translator's Note.*

Examined from the outside with the lens, the black parts reveal nothing out of the common. The black is homogeneous and everywhere of equal depth. On the other hand, in the coloured portions, we see little polygonal, granular masses, forming a close-meshed network. By cutting round the circumference of the abdomen with a pair of scissors, the horny integument of the dorsal surface may readily be removed in one piece, without any shreds of the organs which it protected. This large strip of skin is transparent in the zones that correspond with the white bands in the natural state; it is black or yellow on the black or yellow bands. These last indeed owe their colouring to a layer of pigment which the point of a paintbrush will easily loosen and remove.

As for the white bands, their origin is this: once the skin has been removed, the dorsal surface of the abdomen, whose graceful mosaic is not in any way disturbed, reveals a layer of polygonal white spots, distributed in belts, here densely and there less so. The denser belts correspond with the white bands. It is their magnificent opaque white granulations which, seen through the transparent skin, form the snow-white stripes in the live Spider.

Treated with nitric acid on the microscopic slide, they do not dissolve nor produce effervescence. Uric acid then is not present in this case; and the substance must be guanine, an alkaloid known to be the urinary product of the Spiders. The same is true of the yellow, black, purple or orange pigment that forms a coating under the skin. In short, by utilizing, in a different chemical combination, the waste products of animal oxidization, the magnificent Spider rivals the magnificent caterpillar; she beautifies herself with guanine as the other does with its uric acid.

Let us abridge this dry subject; let us be content with these few data, which could if necessary be corroborated by many others. What does the little that we have learnt teach us? It tells us that the materials rejected by the organism, guanine, uric acid and other dross from life's refinery, play an important part in the coloration of the insect.

The Glow-Worm and Other Beetles

Two cases are distinguishable, according as the colour is dyed or simply painted. The skin, itself colourless and transparent, is in places illumined by a coloured varnish, which can be removed by a touch with a paintbrush. Here we have paint, the result of the urinary compound laid on the inner surface of the covering, just as the chromatic ingredients of our glass-painters are laid on our stained-glass windows.

At other places the skin is coloured in its very substance; the colouring-matter forms an integral part of it and can no longer be swept away with a camel-hair brush. Here we have a dyed fabric, represented in our windows by the panes of coloured glass which the crucible decorates uniformly with this or that tint, by means of the incorporated metallic oxides.

Whereas, in these two cases, there is a profound difference in the distribution of the chromatic materials, is this true of their chemical nature as well? The suggestion is hardly admissible. The worker in stained glass dyes or paints with the same oxides. Life, that incomparable artist, must even more readily obtain an infinite variety of results by uniformity of method.

It shows us, on the back of the Spurge Caterpillar,[7] black spots jumbled up with other spots, white, yellow or red. Paints and dyes lie side by side. Is there on this side of the dividing line a paint-stuff and on the other side a dye-stuff, absolutely different in character from the first? While chemistry is not yet in a position to demonstrate, with its reagents, the common origin of the two substances, at least the most convincing analogies point to it.

[7] The caterpillar of the Spurge Hawk-moth.—*Translator's Note*.

In this delicate problem of the insect's colouring, one single point thus far comes within the domain of observed facts: the progressive advance of chromatic evolution. The carbuncle of the Dung-Beetle of the Pampas suggested the question. Let us then inquire of his near neighbours, who will perhaps enable us to advance a step farther.

Newly stripped of his cast-off nymphal skin, the Sacred Beetle possesses a strange costume, bearing no resemblance to the ebony black which will be the portion of the mature insect. The head, legs and thorax are a bright rusty red; the wing-cases and abdomen are white. As a colour, the red is almost that of the Spurge Caterpillar, but it is the result of a dye on which nitric acid has no effect as a detector of urates. The same chromatic principle must certainly exist in a more elaborate form and under a different molecular arrangement in the skin of the abdomen and the wing-cases which will presently replace white by red.

In two or three days the colourless becomes the coloured, a process whose rapidity implies a fresh molecular structure rather than a change of composition. The building-stone remains the same, but is arranged in a different order; and the structure alters in appearance.

The Scarabæus is now all red. The first brown stains show themselves on the denticulations of the forehead and fore-legs, the sign of an earlier maturity in the implements of labour, which are to acquire an exceptional hardness. The smoky tinge spreads more or less all over the insect, replaces the red, turns darker and finally becomes the regulation black. In less than a week the colourless insect turns a rusty red, next a sooty brown and then an ebony black. The process is completed; the insect possesses its normal colouring.

Even so do the Copres, the Gymnopleuri,[8] the Onites, the Onthophagi and many others behave; even so must the jewel of the pampas, the Splendid Phanæus set to work. With as much certainty as though I had him before my eyes at the moment when he divests himself of his nymphal swaddling-bands, I see him a dull red, rusty or crimson, excepting on the wing-covers and the abdomen, which are at first colourless and presently turn the same colour as the rest. In the Sacred Beetle this initial red is followed by black; the Phanæus replaces it by the brilliance of copper and the reflections of the emerald. Ebony, metal, the gem: have they the same origin here then? Evidently.

[8] Cf. *The Sacred Beetle and Others:* chap. viii.—*Translator's Note.*

The metallic lustre does not call for a change of nature; a mere nothing is enough to produce it. Silver, when very finely subdivided by the methods whereof chemistry knows the secret, becomes a dust as poor to look at as soot. When pressed between two hard bodies, this dirty powder, which might be dried mud, at once acquires the metallic sheen and again becomes the silver which we know. A mere molecular contact has wrought the miracle.

Dissolved in water, the murexide derived from uric acid is a magnificent crimson. Solidified by crystallization, it rivals in splendour the gold-green of the Cantharides. The widely-used fuschine affords a well-known example of like properties.

Everything, then, appears to show that the same substance, derived from urinary excretions, yields, according to the mode in which its ultimate particles are grouped, the metallic red of the Phanæus, as well as the white, the dull red and the black of the Sacred Beetle. It becomes black on the dorsal surface of the Stercoraceous Geotrupes and the Mimic Geotrupes; and, with a quick change, it turns into amethyst under the belly of the first and into copper pyrites under the belly of the second. It covers the back of *Cetonia floricola* with golden bronze and the under surface with metallic purple. According to the insect, according to the part of the body, it remains a dingy compound or sparkles with reflections even more vivid and varied than those possessed by the metals.

Light seems irrelevant to the development of these splendours; it neither accelerates nor retards them. Since direct exposure to the sun, owing to the excess of heat, is fatal to the delicate process of the nymphosis, I shaded the solar rays with a screen of water contained between slips of glass; and to the bright light thus moderated in temperature I daily, throughout the period of chromatic evolution, subjected a number of Sacred Beetles, Geotrupes and Cetoniæ. As standards of comparison I had witnesses of whom I kept some in diffused light and others in complete darkness. My experiments had no appreciable result. The development of the colours took place in the sunlight and in the dark alike, neither more rapidly nor more slowly and without difference in the tints.

The Glow-Worm and Other Beetles

This negative result was easy to foresee. The Buprestis emerging from the depths of the trunk in which he has spent his larval life; the Geotrupes and the Phanæus leaving their natal burrows possess their final adornments, which will not become richer in the rays of the sun, at the time when they make their appearance in the open air. The insect does not claim the assistance of the light for its colour chemistry, not even the Cicada,[9] who bursts her larval scabbard and changes from pale green to brown as easily in the darkness of my apparatus as in the sunlight, in the usual manner.

[9] Cf. *The Life of the Grasshopper:* chaps. i. to v. — *Translator's Note.*

The chromatics of the insect, having as its basis the urinary waste products, might well be found in various animals of a higher order. We know of at least one example. The pigment of a small American lizard is converted into uric acid under the prolonged action of boiling hydrochloric acid.[10] This cannot be an isolated instance; and there is reason to believe that the reptilian class daubs its garments with similar products.

[10] A. B. Griffiths, Transactions of the Académie des sciences, 26 November, 1894. — *Author's Note.*

From the reptile to the bird is no great distance. Then the Wood-pigeon's iridescent hues, the eyes on the Peacock's tail, the Kingfisher's sea-blue, the Flamingo's carmine are more or less closely connected with the urinary excretions? Why not? Nature, that sublime economist, delights in these vast antitheses which upset all our conceptions of the values of things. Of a pinch of common charcoal she makes a diamond; of the same clay which the potter fashions into a bowl for the Cat's supper she makes a ruby; of the filthy waste products of the organism she makes the splendours of the insect and the bird. The metallic marvels of the Buprestis and the Ground-beetle; the amethyst, ruby, sapphire, emerald and topaz of the Humming-bird; glories which would exhaust the language of the lapidary jeweller: what are they in reality? Answer: a drop of urine.

CHAPTER XI
THE BURYING-BEETLES: THE BURIAL

Beside the footpath in April lies the Mole, disembowelled by the peasant's spade; at the foot of the hedge the pitiless urchin has stoned to death the Lizard, who was about to don his green, pearl-embellished costume. The passer-by has thought it a meritorious deed to crush beneath his heel the chance-met Adder; and a gust of wind has thrown a tiny unfledged bird from its nest. What will become of these little bodies and so many other pitiful remnants of life? They will not long offend our sense of sight and smell. The sanitary officers of the fields are legion.

An eager freebooter, ready for any task, the Ant is the first to come hastening and begin, particle by particle, to dissect the corpse. Soon the odour attracts the Fly, the genitrix of the odious maggot. At the same time, the flattened Silpha,[1] the glistening, slow-trotting Cellar-beetle, the Dermestes,[2] powdered with snow upon the abdomen, and the slender Staphylinus,[3] all, whence coming no one knows, hurry hither in squads, with never-wearied zeal, investigating, probing and draining the infection.

[1] Or Carrion-beetle. — *Translator's Note.*

[2] Or Bacon-beetle. — *Translator's Note.*

[3] Or Rove-beetle. — *Translator's Note.*

What a spectacle, in the spring, beneath a dead Mole! The horror of this laboratory is a beautiful sight for one who is able to observe and to meditate. Let us overcome our disgust; let us turn over the unclean refuse with our foot. What a swarming there is beneath it, what a tumult of busy workers! The Silphæ, with wing-cases wide and dark, as though in mourning, flee distraught, hiding in the cracks in the soil; the Saprini,[4] of polished ebony which mirrors the sunlight, jog hastily off, deserting their workshop; the Dermestes, of whom one wears a fawn-coloured tippet flecked with white, seek to fly away, but, tipsy with the putrid nectar, tumble over and reveal the immaculate whiteness of their bellies, which forms a violent contrast with the gloom of the rest of their attire.

[4] The Saprinus is a very small carnivorous Beetle. Cf. *The Life of the Fly:* chap. xvi.—*Translator's Note.*

What were they doing there, all these feverish workers? They were making a clearance of death on behalf of life. Transcendent alchemists, they were transforming that horrible putrescence into a living and inoffensive product. They were draining the dangerous corpse to the point of rendering it as dry and sonorous as the remains of an old slipper hardened on the refuse-heap by the frosts of winter and the heats of summer. They were working their hardest to render the carrion innocuous.

Others will soon put in their appearance, smaller creatures and more patient, who will take over the relic and exploit it ligament by ligament, bone by bone, hair by hair, until the whole has been restored to the treasury of life. All honour to these purifiers! Let us put back the Mole and go our way.

Some other victim of the agricultural labours of spring, a Shrew-mouse, Field-mouse, Mole, Frog, Adder, or Lizard, will provide us with the most vigorous and famous of these expurgators of the soil. This is the Burying-beetle, the Necrophorus, so different from the cadaveric mob in dress and habits. In honour of his exalted functions he exhales an odour of musk; he bears a red tuft at the tip of his antennæ; his breast is covered with nankeen; and across his wing-cases he wears a double, scalloped scarf of vermillion. An elegant, almost sumptuous costume, very superior to that of the others, but yet lugubrious, as befits your undertaker's man.

He is no anatomical dissector, cutting his subject open, carving its flesh with the scalpel of his mandibles; he is literally a grave-digger, a sexton. While the others—Silphæ, Dermestes, Cellar-beetles—gorge themselves with the exploited flesh, without, of course, forgetting the interests of the family, he, a frugal eater, hardly touches his find on his own account. He buries it entire, on the spot, in a cellar where the thing, duly ripened, will form the diet of his larvæ. He buries it in order to establish his progeny.

This hoarder of dead bodies, with his stiff and almost heavy movements, is astonishingly quick at storing away wreckage. In a shift of a few hours, a comparatively enormous animal, a Mole, for instance, disappears, engulfed by the earth. The others leave the dried, emptied carcass to the air, the sport of the winds for months on end; he, treating it as a whole, makes a clean job of things at once. No visible trace of his work remains but a tiny hillock, a burial-mound, a tumulus.

With his expeditious method, the Necrophorus is the first of the little purifiers of the fields. He is also one of the most celebrated of insects in respect of his psychical capacities. This undertaker is endowed, they say, with intellectual faculties approaching to reason, such as are not possessed by the most gifted of the Bees and Wasps, the collectors of honey or game. He is honoured by the two following anecdotes, which I quote from Lacordaire's[5] *Introduction a l'entomologie*, the only general treatise at my disposal:

"Clairville," says the author, "reports that he saw a *Necrophorus vespillo*, who, wishing to bury a dead Mouse and finding the soil on which the body lay too hard, went to dig a hole at some distance, in soil more easily displaced. This operation completed, he attempted to bury the Mouse in the cavity, but, not succeeding, he flew away and returned a few moments later, accompanied by four of his fellows, who assisted him to move the Mouse and bury it."

[5] Jean Théodore Lacordaire (1801-1870), author of *Genera des coléoptères* (1854-1876) and of the work quoted above (1837-1839).—*Translator's Note.*

In such actions, Lacordaire adds, we cannot refuse to admit the intervention of reason.

"The following case," he continues, "recorded by Gleditsch,[6] has also every indication of the intervention of reason. One of his friends, wishing to desiccate a Frog, placed it on the top of a stick thrust into the ground, in order to make sure that the Necrophori should not come and carry it off. But this precaution was of no effect; the insects, being unable to reach the Frog, dug under the stick and, having caused it to fall, buried it as well as the body."[7]

[6] Johann Gottlieb Gleditsch (1714-1786), the German botanist.— *Translator's Note.*

[7] *Suites à Buffon. Introduction a l'entomologie*, vol. ii., pp. 460-61.— *Author's Note.*

To grant, in the intellect of the insect, a lucid understanding of the relations between cause and effect, between the end and the means, is to make a statement of serious import. I know of scarcely any more suited to the philosophical brutalities of my time. But are these two anecdotes really true? Do they involve the consequences deduced from them? Are not those who accept them as sound evidence just a little too simple?

To be sure, simplicity is needed in entomology. Without a good dose of this quality, a mental defect in the eyes of practical folk, who would busy himself with the lesser creatures? Yes, let us be simple, without being childishly credulous. Before making insects reason, let us reason a little ourselves; let us, above all, consult the experimental test. A fact gathered at random, without criticism, cannot establish a law.

I do not propose, O valiant grave-diggers, to depreciate your merits; such is far from being my intention. I have that in my notes, on the other hand, which will do you more honour than the story of the gibbet and the Frog; I have gleaned, for your benefit, examples of prowess which will shed a new lustre upon your reputation.

No, my intention is not to belittle your renown. Besides, it is not the business of impartial history to maintain a given thesis; it follows facts. I wish simply to question you upon the power of logic attributed to you. Do you or do you not enjoy gleams of reason? Have you within you the humble germ of human thought? That is the problem before us.

To solve it we will not rely upon the accidents which good fortune may now and again procure for us. We must employ the breeding-cage, which will permit of assiduous visits, continuous enquiry and a variety of artifices. But how to stock the cage? The land of the olive-tree is not rich in Necrophori. To my knowledge it possesses only a single species, *N. vestigator*, HERSCH.; and even this rival of the grave-diggers of the north is pretty scarce. The discovery of three or four in the spring was as much as my hunting-expeditions yielded in the old days. This time, if I do not resort to the ruses of the trapper, I shall obtain no more than that, whereas I stand in need of at least a dozen.

These ruses are very simple. To go in search of the sexton, who exists only here and there in the country-side, would be nearly always a waste of time; the favourable month, April, would be past before my cage was suitably stocked. To run after him is to trust too much to accident; so we will make him come to us by scattering in the orchard an abundant collection of dead Moles. To this carrion, ripened by the sun, the insect will not fail to hasten from the various points of the horizon, so accomplished is he in detecting such a delicacy.

I make an arrangement with a gardener in the neighbourhood, who, two or three times a week, makes up for the penury of my two acres of stony ground by providing me with vegetables raised in a better soil. I explain to him my urgent need of Moles in unlimited numbers. Battling daily with trap and spade against the importunate excavator who uproots his crops, he is in a better position than any one to procure for me what I regard for the moment as more precious than his bunches of asparagus or his white-heart cabbages.

The worthy man at first laughs at my request, being greatly surprised by the importance which I attribute to the abhorrent animal, the *Darboun;* but at last he consents, not without a suspicion at the back of his mind that I am going to make myself a gorgeous winter waist-coat with the soft, velvety skins of the Moles. A thing like that must be good for pains in the back. Very well. We settle the matter. The essential thing is that the *Darbouns* reach me.

They reach me punctually, by twos, by threes, by fours, packed in a few cabbage-leaves, at the bottom of the gardener's basket. The excellent fellow who lent himself with such good grace to my strange wishes will never guess how much comparative psychology will owe him! In a few days I was the possessor of thirty Moles, which were scattered here and there, as they reached me, in bare spots of the orchard, among the rosemary-bushes, the strawberry-trees and the lavender-beds.

Now it only remained to wait and to examine, several times a day, the under-side of my little corpses, a disgusting task which any one would avoid whose veins were not filled with the sacred fire of enthusiasm. Only little Paul, of all the household, lent me the aid of his nimble hand to seize the fugitives. I have already said that the entomologist needs simplicity of mind. In this important business of the Necrophori, my assistants were a small boy and an illiterate.

Little Paul's visits alternating with mine, we had not long to wait. The four winds of heaven bore forth in all directions the odour of the carrion; and the undertakers hurried up, so that the experiments, begun with four subjects, were continued with fourteen, a number not attained during the whole of my previous searches, which were unpremeditated and in which no bait was used as decoy. My trapper's ruse was completely successful.

Before I report the results obtained in the cage, let us stop for a moment to consider the normal conditions of the labours that fall to the lot of the Necrophori. The Beetle does not select his head of game, choosing one in proportion to his strength, as do the Hunting Wasps; he accepts what chance offers. Among his finds some are small, such as the Shrew-mouse; some medium-sized, such as the Field-mouse; some enormous, such as the Mole, the Sewer-rat and the Snake, any of which exceeds the digging-powers of a single sexton. In the majority of cases, transportation is impossible, so greatly disproportioned is the burden to the motive-power. A slight displacement, caused by the effort of the insects' backs, is all that can possibly be effected.

Ammophila and Cerceris,[8] Sphex and Pompilus excavate their burrows wherever they please; they carry their prey on the wing, or, if too heavy, drag it afoot. The Necrophorus knows no such facilities in his task. Incapable of carting the monstrous corpse, no matter where encountered, he is forced to dig the grave where the body lies.

[8] Cf. *The Hunting Wasps:* chaps. i. to iii.—*Translator's Note.*

This obligatory place of sepulture may be in stony soil or in shifting sand; it may occupy this or that bare spot, or some other where the grass, especially the couch-grass, plunges into the ground its inextricable network of little cords. There is a great probability, too, that a bristle of stunted brambles may be supporting the body at some inches above the soil. Slung by the labourer's spade, which has just broken his back, the Mole falls here, there, anywhere, at random; and where the body falls, no matter what the obstacles, provided that they be not insurmountable, there the undertaker must utilize it.

The difficulties of inhumation are capable of such variety as causes us already to foresee that the Necrophorus cannot employ fixed methods in performing his task. Exposed to fortuitous hazards, he must be able to modify his tactics within the limits of his modest discernment. To saw, to break, to disentangle, to lift, to shake, to displace: these are so many means which are indispensable to the grave-digger in a predicament. Deprived of these resources, reduced to uniformity of procedure, the insect would be incapable of pursuing its calling.

We see at once how imprudent it would be to draw conclusions from an isolated case in which rational co-ordination or premeditated intention might appear to play its part. Every instinctive action no doubt has its motive; but does the animal in the first place judge whether the action is opportune? Let us begin by a careful consideration of the creature's labours; let us support each piece of evidence by others; and then we shall perhaps be able to answer the question.

First of all, a word as to diet. A general scavenger, the Burying-beetle refuses no sort of cadaveric putrescence. All is good to his senses, feathered game or furry, provided that the burden do not exceed his strength. He exploits the batrachian or the reptile with no less animation. He accepts without hesitation extraordinary finds, probably unknown to his race, as witness a certain Goldfish, a red Chinese Carp, whose body, placed in one of my cages, was forthwith considered an excellent tit-bit and buried according to the rules. Nor is butcher's meat despised. A mutton-cutlet, a strip of beef-steak, in the right stage of maturity, disappeared beneath the soil, receiving the same attentions as those lavished on the Mole or the Mouse. In short, the Necrophorus has no exclusive preferences; anything putrid he conveys underground.

The maintenance of his industry, therefore, presents no sort of difficulty. If one kind of game be lacking, some other, the first to hand, will very well replace it. Nor is there much trouble in fixing the site of his industry. A capacious wire-gauze cover, resting on an earthen pan filled to the brim with fresh, heaped sand, is sufficient. To obviate criminal attempts on the part of the Cats, whom the game would not fail to tempt, the cage is installed in a closed glass-house, which in winter shelters the plants and in summer serves as an entomological laboratory.

Now to work. The Mole lies in the centre of the enclosure. The soil, easily shifted and homogeneous, realizes the best conditions for comfortable work. Four Necrophori, three males and a female, are there with the body. They remain invisible, hidden beneath the carcase, which from time to time seems to return to life, shaken from end to end by the backs of the workers. An observer not in the secret would be somewhat astonished to see the dead creature move. From time to time, one of the sextons, almost always a male, comes out and walks round the animal, which he explores, probing its velvet coat. He hurriedly returns, appears again, once more investigates and creeps back under the corpse.

The tremors become more pronounced; the carcase oscillates, while a cushion of sand, pushed out from below, grows up all around it. The Mole, by reason of his own weight and the efforts of the grave-diggers, who are labouring at their task underneath, gradually sinks, for lack of support, into the undermined soil.

Presently the sand which has been pushed out quivers under the thrust of the invisible miners, slips into the pit and covers the interred Mole. It is a clandestine burial. The body seems to disappear of itself, as though engulfed by a fluid medium. For a long time yet, until the depth is regarded as sufficient, the body will continue to descend.

It is, on the whole, a very simple operation. As the diggers below deepen the cavity into which the corpse, shaken and tugged above, sinks without the direct intervention of the sextons, the grave fills of itself by the mere slipping of the soil. Stout shovels at the tips of their claws, powerful backs, capable of creating a little earthquake: the diggers need nothing more for the practice of their profession. Let us add—for this is an essential point—the art of continually jerking the body, so as to pack it into a lesser volume and make it glide through difficult passages. We shall soon see that this art plays a leading part in the industry of the Necrophori.

Although he has disappeared, the Mole is still far from having reached his destination. Let us leave the undertakers to finish their job. What they are now doing below ground is a continuation of what they did on the surface and would teach us nothing new. We will wait for two or three days.

The moment has come. Let us inform ourselves of what is happening down there. Let us visit the place of corruption. I shall never invite anybody to the exhumation. Of those about me, only little Paul has the courage to assist me.

The Mole is a Mole no longer, but a greenish horror, putrid, hairless, shrunk into a sort of fat, greasy rasher. The thing must have undergone careful manipulation to be thus condensed into a small volume, like a fowl in the hands of the cook, and, above all, to be so completely deprived of its furry coat. Is this culinary procedure undertaken in respect of the larvæ, which might be incommoded by the fur? Or is it just a casual result, a mere loss of hair due to putridity? I am not certain. But it is always the case that these exhumations, from first to last, have revealed the furry game furless and the feathered game featherless, except for the pinion- and tail-feathers. Reptiles and fish, on the other hand, retain their scales.

Let us return to the unrecognizable thing that was once a Mole. The tit-bit lies in a spacious crypt, with firm walls, a regular workshop, worthy of being the bake-house of a Copris. Except for the fur, which lies scattered about in flocks, it is intact. The grave-diggers have not eaten into it: it is the patrimony of the sons, not the provision of the parents, who, to sustain themselves, levy at most a few mouthfuls of the ooze of putrid humours.

Beside the dish which they are kneading and protecting are two Necrophori; a couple, no more. Four collaborated in the burial. What has become of the other two, both males? I find them hidden in the soil, at a distance, almost on the surface.

This observation is not an isolated one. Whenever I am present at a funeral undertaken by a squad in which the males, zealous one and all, predominate, I find presently, when the burial is completed, only one couple in the mortuary cellar. After lending their assistance, the rest have discreetly retired.

These grave-diggers, in truth, are remarkable fathers. They have nothing of the happy-go-lucky paternal carelessness that is the general rule among insects, which pester the mother for a moment with their attentions and then leave her to care for the offspring! But those who would be idlers in the other castes here labour valiantly, now in the interest of their own family, now in that of another's, without distinction. If a couple is in difficulties, helpers arrive, attracted by the odour of carrion; anxious to serve a lady, they creep under the body, work at it with back and claw, bury it and then go their ways, leaving the master and mistress of the house to their happiness.

For some time longer these two manipulate the morsel in concert, stripping it of fur or feather, trussing it and allowing it to simmer to the grub's taste. When everything is in order, the couple go forth, dissolving their partnership; and each, following his fancy, begins again elsewhere, even if only as a mere auxiliary.

Twice and no oftener hitherto have I found the father preoccupied by the future of his sons and labouring in order to leave them rich: it happens with certain dung-workers and with the Necrophori, who bury dead bodies. Scavengers and undertakers both have exemplary morals. Who would look for virtue in such a quarter?

What follows—the larval existence and the metamorphosis—is a secondary and, for that matter, a familiar detail. It is a dry subject and I will deal with it briefly. At the end of May, I exhume a Brown Rat, buried by the grave-diggers a fortnight earlier. Transformed into a black, sticky mass, the horrible dish provides me with fifteen larvæ already, for the most part, of the normal size. A few adults, unquestionably connections of the brood, are also swarming amid the putrescence. The laying-time is over now and victuals are plentiful. Having nothing else to do, the foster-parents have sat down to the feast with the nurslings.

The undertakers are quick at rearing a family. It is at most a fortnight since the Rat was laid in the earth; and here already is a vigorous population on the verge of the metamorphosis. This precocity amazes me. It would seem as though carrion liquefaction, deadly to any other stomach, were in this case a food productive of special energy, which stimulates the organism and accelerates its growth, so that the fare may be consumed before its approaching conversion into mould. Living chemistry makes haste to outstrip the ultimate reactions of mineral chemistry.

The Glow-Worm and Other Beetles

White, naked, blind, possessing the customary attributes of life spent in the dark, the larva, with its tapering outline, is slightly reminiscent of the Ground-beetles'. The mandibles are black and powerful and make excellent dissecting-scissors. The limbs are short, but capable of a quick, toddling gait. The segments of the abdomen are clad on the upper surface in a narrow red plate, armed with four little spikes, whose office apparently is to furnish points of support when the larva quits the natal dwelling and dives into the soil, there to undergo the transformation. The thoracic segments are provided with wider plates, but unarmed.

The adults discovered in the company of their larval family, in this putrescence which was a Rat, are all abominably verminous. So shiny and neat in their attire, when at work under the first Moles of April, the Necrophori, when June approaches, become odious to look upon. A layer of parasites envelops them; insinuating itself into the joints, it forms an almost continuous crust. The insect presents a misshapen appearance under this overcoat of vermin, which my hair-pencil can hardly brush aside. Driven off the belly, the horde runs round the sufferer, perches on his back and refuses to let go.

I recognize the Beetle's Gamasus, the Tick who so often soils the ventral amethyst of our Geotrupes. No, life's prizes do not go to the useful. Necrophori and Geotrupes devote themselves to the general health; and these two corporations, so interesting in their hygienic functions, so remarkable for their domestic morals, fall victims to the vermin of poverty. Alas, of this discrepancy between the services rendered and the harshness of life there are many other examples outside the world of scavengers and undertakers!

The Burying-beetles display an exemplary domestic morality, but it does not continue till the end. In the first fortnight of June, the family being sufficiently provided, the sextons strike work and my cages are deserted on the surface, in spite of new arrivals of Mice and Sparrows. From time to time, some grave-digger leaves the subsoil and comes crawling languidly into the fresh air.

The Glow-Worm and Other Beetles

Another rather curious fact now attracts my attention. All those who climb up from underground are maimed, with limbs amputated at the joints, some higher up, some lower down. I see one cripple who has only one leg left entire. With this odd limb and the stumps of the others, lamentably tattered, scaly with vermin, he rows, as it were, over the sheet of dust. A comrade emerges, better off for legs, who finishes the invalid and cleans out his abdomen. Thus do my thirteen remaining Necrophori end their days, half-devoured by their companions, or at least shorn of several limbs. The pacific relations of the outset are succeeded by cannibalism.

History tells us that certain peoples, the Massagetæ and others, used to kill off their old men to save them from senile misery. The fatal blow on the hoary skull was in their eyes an act of filial piety. The Necrophori have their share of these ancient barbarities. Full of days and henceforth useless, dragging out a weary existence, they mutually exterminate one another. Why prolong the agony of the impotent and the imbecile?

The Massagetæ might plead, as an excuse for their atrocious custom, a dearth of provisions, which is an evil counsellor; not so the Necrophori, for, thanks to my generosity, victuals are more than plentiful, both beneath the soil and on the surface. Famine plays no part in this slaughter. What we see is an aberration due to exhaustion, the morbid fury of a life on the point of extinction. As is generally the case, work bestows a peaceable disposition on the grave-digger, while inaction inspires him with perverted tastes. Having nothing left to do, he breaks his kinsman's limbs and eats him up, heedless of being maimed or eaten himself. It is the final deliverance of verminous old age.

This murderous frenzy, breaking out late in life, is not peculiar to the Necrophorus. I have described elsewhere the perversity of the Osmia, so placid in the beginning. Feeling her ovaries exhausted, she smashes her neighbours' cells and even her own; she scatters the dusty honey, rips open the egg, eats it. The Mantis devours the lovers who have played their parts; the mother Decticus willingly nibbles a thigh of her decrepit husband; the merry Crickets, once the eggs are laid in the ground, indulge in tragic domestic quarrels and with not the least compunction slash open one another's bellies. When the cares of the family are finished, the joys of life are finished likewise. The insect then sometimes becomes depraved; and its disordered mechanism ends in aberrations.

The larva has nothing striking to show in the way of industry. When it has fattened to the desired extent, it leaves the charnel-house of the natal crypt and descends into the earth, far from the putrefaction. Here, working with its legs and its dorsal armour, it presses back the sand around it and makes itself a close cabin wherein to rest for the metamorphosis. When the lodge is ready and the torpor of the approaching moult arrives, it lies inert; but, at the least alarm, it comes to life and turns round on its axis.

Even so do several nymphs spin round and round when disturbed, notably that of *Ægosomus scabricornis* which I have now before my eyes in July. It is always a fresh surprise to see these mummies suddenly throw off their immobility and gyrate on their own axis with a mechanism whose secret deserves to be fathomed. The science of rational mechanics might find something here to whet its finest theories upon. The strength and litheness of a clown cannot compare with those of this budding flesh, this hardly coagulated glair.

Once isolated in its cell, the larva of the Necrophorus becomes a nymph in ten days or so. I lack the evidence furnished by direct observation, but the story is completed of itself. The Necrophorus must assume the adult form in the course of the summer; like the Dung-beetle, he must enjoy in the autumn a few days of revelry free from family cares. Then, when the cold weather draws near, he goes to earth in his winter quarters, whence he emerges as soon as spring arrives.

CHAPTER XII
THE BURYING-BEETLES: EXPERIMENTS

Let us come to the feats of reason which have earned for the Necrophorus the best part of his fame and, to begin with, submit the case related by Clairville, that of the too hard soil and the call for assistance, to the test of experiment.

With this object I pave the centre of the space beneath the cover, flush with the soil, with a brick, which I sprinkle with a thin layer of sand. This will be the soil that cannot be dug. All around it, for some distance and on the same level, lies the loose soil, which is easy to delve.

In order to approach the conditions of the anecdote, I must have a Mouse; with a Mole, a heavy mass, the removal would perhaps present too much difficulty. To obtain one, I place my friends and neighbours under requisition; they laugh at my whim but none the less proffer their traps. Yet, the moment a very common thing is needed, it becomes rare. Defying decency in his speech, after the manner of his ancestors' Latin, the Provençal says, but even more crudely than in my translation:

"If you look for dung, the Donkeys become constipated!"

At last I possess the Mouse of my dreams! She comes to me from that refuge, furnished with a truss of straw, in which official charity grants a day's hospitality to the pauper wandering over the face of the fertile earth, from that municipal hostel whence one inevitably issues covered with Lice. O Réaumur,[1] who used to invite marchionesses to see your caterpillars change their skins, what would you have said of a future disciple conversant with such squalor as this? Perhaps it is well that we should not be ignorant of it, so that we may have compassion with that of the beast.

[1] René Antoine Ferchault de Réaumur (1683-1757), the inventor of the Réaumur thermometer and author of *Mémoires pour servir à l'histoire naturelle des insectes* (1734-1742). — *Translator's Note.*

The Mouse so greatly desired is mine. I place her upon the centre of the brick. The grave-diggers under the wire cover are now seven in number, including three females. All have gone to earth; some are inactive, close to the surface; the rest are busy in their crypts. The presence of the fresh corpse is soon perceived. About seven o'clock in the morning, three Necrophori come hurrying up, two males and a female. They slip under the Mouse, who moves in jerks, a sign of the efforts of the burying-party. An attempt is made to dig into the layer of sand which hides the brick, so that a bank of rubbish accumulates round the body.

For a couple of hours the jerks continue without results. I profit by the circumstance to learn the manner in which the work is performed. The bare brick allows me to see what the excavated soil would conceal from me. When it is necessary to move the body, the Beetle turns over; with his six claws he grips the hair of the dead animal, props himself upon his back and pushes, using his forehead and the tip of his abdomen as a lever. When he wants to dig, he resumes the normal position. So, turn and turn about, the sexton strives, now with his legs in the air, when it is a question of shifting the body or dragging it lower down; now with his feet on the ground, when it is necessary to enlarge the grave.

The point at which the Mouse lies is finally recognized as unassailable. A male appears in the open. He explores the corpse, goes round it, scratches a little at random. He goes back; and immediately the dead body rocks. Is he advising his collaborators of what he has discovered? Is he arranging the work with a view to their establishing themselves elsewhere, on propitious soil?

The facts are far from confirming this idea. When he shakes the body, the others imitate him and push, but without combining their efforts in a given direction, for, after advancing a little towards the edge of the brick, the burden goes back again, returning to the point of departure. In the absence of a concerted understanding, their efforts of leverage are wasted. Nearly three hours are occupied by oscillations which mutually annul one another. The Mouse does not cross the little sand-hill heaped about her by the rakes of the workers.

For the second time, a male appears and makes a round of exploration. A boring is effected in loose earth, close beside the brick. This is a trial excavation, to learn the nature of the soil, a narrow well, of no great depth, into which the insect plunges to half its length. The well-sinker returns to the other workers, who arch their backs, and the load progresses a finger's-breadth towards the point recognized as favourable. Have we done the trick this time? No, for after a while the Mouse recoils. There is no progress towards a solution of the difficulty.

Now two males come out in search of information, each of his own accord. Instead of stopping at the point already sounded, a point most judiciously chosen, it seemed, on account of its proximity, which would save laborious carting, they precipitately scour the whole area of the cage, trying the soil on this side and on that and ploughing superficial furrows in it. They get as far from the brick as the limits of the enclosure permit.

They dig, by preference, against the base of the cover; here they make several borings, without any reason, so far as I can see, the bed of soil being everywhere equally assailable away from the brick; the first point sounded is abandoned for a second, which is rejected in its turn. A third and fourth are tried; then another. At the sixth point the choice is made. In all these cases the excavation is by no means a grave destined to receive the Mouse, but a mere trial boring, of inconsiderable depth and of the diameter of the digger's body.

Back again to the Mouse, who suddenly shakes, swings, advances, recoils, first in one direction, then in another, until in the end the hillock of sand is crossed. Now we are free of the brick and on excellent soil. Little by little the load advances. This is no cartage by a team hauling in the open, but a jerky removal, the work of invisible levers. The body seems to shift of its own accord.

This time, after all those hesitations, the efforts are concerted; at least, the load reaches the region sounded far more rapidly than I expected. Then begins the burial, according to the usual method. It is one o'clock. It has taken the Necrophori halfway round the clock to ascertain the condition of the locality and to displace the Mouse.

In this experiment it appears, in the first place, that the males play a major part in the affairs of the household. Better-equipped, perhaps, than their mates, they make investigations when a difficulty occurs; they inspect the soil, recognize whence the check arises and choose the spot at which the grave shall be dug. In the lengthy experiment of the brick, the two males alone explored the surroundings and set to work to solve the difficulty. Trusting her assistants, the female, motionless beneath the Mouse, awaited the result of their enquiries. The tests which are to follow will confirm the merits of these valiant auxiliaries.

In the second place, the points where the Mouse lies being recognized as presenting an insurmountable resistance, there is no grave dug in advance, a little farther off, in the loose soil. All the attempts are limited, I repeat, to shallow soundings, which inform the insect of the possibility of inhumation.

It is absolute nonsense to speak of their first preparing the grave to which the body will afterwards be carted. In order to excavate the soil, our sextons have to feel the weight of their dead upon their backs. They work only when stimulated by the contact of its fur. Never, never in this world, do they venture to dig a grave unless the body to be buried already occupies the site of the cavity. This is absolutely confirmed by my two months and more of daily observations.

The rest of Clairville's anecdote bears examination no better. We are told that the Necrophorus in difficulties goes in search of assistance and returns with companions who assist him to bury the Mouse.

This, in another form, is the edifying story of the Sacred Beetle whose pellet has rolled into a rut. Powerless to withdraw his booty from the abyss, the wily Dung-beetle summons three or four of his neighbours, who kindly pull out the pellet and return to their labours when the work of salvage is done.[2]

[2] For the confutation of this theory, cf. *The Sacred Beetle and Others:* chap. i. — *Translator's Note.*

The ill-interpreted exploit of the thieving pill-roller sets me on my guard against that of the undertaker. Shall I be too particular if I ask what precautions the observer took to recognize the owner of the Mouse on his return, when he reappears, as we are told, with four assistants? What sign denotes that one of the five who was able, in so rational a manner, to call for help? Can we even be sure that the one to disappear returns and forms one of the band? There is nothing to tell us so; and this was the essential point which a sterling observer was bound not to neglect. Were they not rather five chance Necrophori who, guided by the smell, without any previous understanding, hastened to the abandoned Mouse to exploit her on their own account? I incline to this opinion, the likeliest of all in the absence of exact information.

Probability becomes certainty if we check the fact by experiment. The test with the brick already tells us something. For six hours my three specimens exhausted themselves in efforts before they succeeded in removing their booty and placing it on practicable soil. In this long and heavy job, helpful neighbours would have been most welcome. Four other Necrophori, buried here and there under a little sand, comrades and acquaintances, fellow-workers of the day before, were occupying the same cage; and not one of the busy ones thought of calling on them to assist. Despite their extreme embarrassment, the owners of the Mouse accomplished their task to the end, without the least help, though this could have been so easily requisitioned.

Being three, one might say, they deemed themselves strong enough; they needed no one else to lend them a hand. The objection does not hold good. On many occasions and under conditions even more difficult than those presented by a hard soil, I have again and again seen isolated Necrophori wearing themselves out against my artifices; yet not once did they leave their workshop to recruit helpers. Collaborators, it is true, often arrive, but they are summoned by their sense of smell, not by the first occupant. They are fortuitous helpers; they are never called in. They are received without strife but also without gratitude. They are not summoned; they are tolerated.

In the glazed shelter where I keep the cage I happened to catch one of these chance assistants in the act. Passing that way in the night and scenting dead flesh, he had entered where none of his kind had yet penetrated of his own accord. I surprised him on the dome of the cover. If the wire had not prevented him, he would have set to work incontinently, in company with the rest. Had my captives invited this one? Assuredly not. Heedless of others' efforts, he hastened up, attracted by the odour of the Mole. So it was with those whose obliging assistance is extolled. I repeat, in respect of their imaginary prowess, what I have said elsewhere of the Sacred Beetle's: it is a child's story, worthy to rank with any fairytale for the amusement of the simple.

A hard soil, necessitating the removal of the body, is not the only difficulty with which the Necrophori are acquainted. Frequently, perhaps more often than not, the ground is covered with grass, above all with couch-grass, whose tenacious rootlets form an inextricable network below the surface. To dig in the interstices is possible, but to drag the dead animal through them is another matter: the meshes of the net are too close to give it passage. Will the grave-digger find himself helpless against such an obstacle, which must be an extremely common one? That could not be.

Exposed to this or that habitual impediment in the exercise of its calling, the animal is always equipped accordingly; otherwise its profession would be impracticable. No end is attained without the necessary means and aptitudes. Besides that of the excavator, the Necrophorus certainly possesses another art: the art of breaking the cables, the roots, the stolons, the slender rhizomes which check the body's descent into the grave. To the work of the shovel and the pick must be added that of the shears. All this is perfectly logical and may be clearly foreseen. Nevertheless, let us call in experiment, the best of witnesses.

I borrow from the kitchen-range an iron trivet whose legs will supply a solid foundation for the engine which I am devising. This is a coarse network made of strips of raffia, a fairly accurate imitation of that of the couch-grass. The very irregular meshes are nowhere wide enough to admit of the passage of the creature to be buried, which this time is a Mole. The machine is planted by its three feet in the soil of the cage, level with the surface. A little sand conceals the ropes. The Mole is placed in the centre; and my bands of sextons are let loose upon the body.

The burial is performed without a hitch in the course of an afternoon. The raffia hammock, almost the equivalent of the natural network of the couch-grass, scarcely disturbs the burying-process. Matters do not proceed quite so quickly; and that is all. No attempt is made to shift the Mole, who sinks into the ground where he lies. When the operation is finished, I remove the trivet. The network is broken at the spot where the corpse was lying. A few strips have been gnawed through; a small number, only as many as were strictly necessary to permit the passage of the body.

Well done, my undertakers! I expected no less of your skill and tact. You foiled the experimenter's wiles by employing the resources which you use against natural obstacles. With mandibles for shears, you patiently cut my strings as you would have gnawed the threads of the grass-roots. This is meritorious, if not deserving of exceptional glorification. The shallowest of the insects that work in earth would have done as much if subjected to similar conditions.

Let us ascend a stage in the series of difficulties. The Mole is now fixed by a strap of raffia fore and aft to a light horizontal cross-bar resting on two firmly-planted forks. It is like a joint of venison on the spit, eccentrically fastened. The dead animal touches the ground throughout the length of its body.

The Necrophori disappear under the corpse and, feeling the contact of its fur, begin to dig. The grave grows deeper and an empty space appears; but the coveted object does not descend, retained as it is by the cross-bar which the two forks keep in place. The digging slackens, the hesitations become prolonged.

However, one of the grave-diggers climbs to the surface, wanders over the Mole, inspects him and ends by perceiving the strap at the back. He gnaws and ravels it tenaciously. I hear the click of the shears that completes the rupture. Crack! The thing is done. Dragged down by his own weight, the Mole sinks into the grave, but slantwise, with his head still outside, kept in place by the second strap.

The Beetles proceed with the burial of the hinder part of the Mole; they twitch and jerk it now in this direction, now in that. Nothing comes of it; the thing refuses to give. A fresh sortie is made by one of them, to find out what is happening overhead. The second strap is perceived, is severed in turn; and henceforth the work goes on as well as could be wished.

My compliments, perspicacious cable-cutters! But I must not exaggerate. The Mole's straps were for you the little cords with which you are so familiar in turfy soil. You broke them, as well as the hammock of the previous experiment, just as you sever with the blades of your shears any natural thread stretching across your catacombs. It is an indispensable trick of your trade. If you had had to learn it by experience, to think it out before practising it, your race would have disappeared, killed by the hesitations of its apprenticeship, for the spots prolific of Moles, Frogs, Lizards and other viands to your taste are usually covered with grass.

You are capable of much better things still; but, before setting forth these, let us examine the case when the ground bristles with slender brushwood, which holds the corpse at a short distance from the ground. Will the find thus hanging where it chances to fall remain unemployed? Will the Necrophori pass on, indifferent to the superb morsel which they see and smell a few inches above their heads, or will they make it drop from its gibbet?

Game does not abound to such a point that it can be despised if a few efforts will obtain it. Before I see the thing happen, I am persuaded that it will fall, that the Necrophori, often confronted with the difficulties of a body not lying on the soil, must possess the instinct to shake it to the ground. The fortuitous support of a few bits of stubble, of a few interlaced twigs, so common in the fields, cannot put them off. The drop of the suspended body, if placed too high, must certainly form part of their instinctive methods. For the rest, let us watch them at work.

I plant in the sand of the cage a meagre tuft of thyme. The shrub is at most some four inches in height. In the branches I place a Mouse, entangling the tail, the paws and the neck among the twigs to increase the difficulty. The population of the cage now consists of fourteen Necrophori and will remain the same until the close of my investigations. Of course they do not all take part simultaneously in the day's work: the majority remain underground, dozing or occupied in setting their cellars in order. Sometimes only one, often two, three or four, rarely more, busy themselves with the corpse which I offer them. To-day, two hasten to the Mouse, who is soon perceived overhead on the tuft of thyme.

They gain the top of the plant by way of the trelliswork of the cage. Here are repeated, with increased hesitation, due to the inconvenient nature of the support, the tactics employed to remove the body when the soil is unfavourable. The insect props itself against a branch, thrusting alternately with back and claws, jerking and shaking vigorously until the point whereat it is working is freed from its fetters. In one brief shift, by dint of heaving their backs, the two collaborators extricate the body from the tangle. Yet another shake; and the Mouse is down. The burial follows.

There is nothing new in this experiment: the find has been treated just as though it lay on soil unsuitable for burial. The fall is the result of an attempt to transport the load.

The Glow-Worm and Other Beetles

The time has come to set up the Frog's gibbet made famous by Gleditsch. The batrachian is not indispensable; a Mole will serve as well or even better. With a ligament of raffia I fix him, by his hind-legs, to a twig which I plant vertically in the ground, inserting it to no great depth. The creature hangs plumb against the gibbet, its head and shoulders making ample contact with the soil.

The grave-diggers set to work beneath the part which lies along the ground, at the very foot of the stake; they dig a funnel into which the Mole's muzzle, head and neck sink little by little. The gibbet becomes uprooted as they descend and ends by falling, dragged over by the weight of its heavy burden. I am assisting at the spectacle of the overturned stake, one of the most astonishing feats of reason with which the insect has ever been credited.

This, for one who is considering the problem of instinct, is an exciting moment. But let us beware of forming conclusions just yet; we might be in too great a hurry. Let us first ask ourselves whether the fall of the stake was intentional or accidental. Did the Necrophori lay it bare with the express purpose of making it fall? Or did they, on the contrary, dig at its base solely in order to bury that part of the Mole which lay on the ground? That is the question, which, for the rest, is very easy to answer.

The experiment is repeated; but this time the gibbet is slanting and the Mole, hanging in a vertical position, touches the ground at a couple of inches from the base of the apparatus. Under these conditions, absolutely no attempt is made to overthrow it. Not the least scrape of a claw is delivered at the foot of the gibbet. The entire work of excavation is performed at a distance, under the body, whose shoulders are lying on the ground. Here and here only a hole is dug to receive the front of the body, the part accessible to the sextons.

A difference of an inch in the position of the suspended animal destroys the famous legend. Even so, many a time, the most elementary sieve, handled with a little logic, is enough to winnow a confused mass of statements and to release the good grain of truth.

Yet another shake of this sieve. The gibbet is slanting or perpendicular, no matter which; but the Mole, fixed by his hind-legs to the top of the twig, does not touch the soil; he hangs a few fingers'-breadths from the ground, out of the sextons' reach.

What will they do now? Will they scrape at the foot of the gibbet in order to overturn it? By no means; and the ingenuous observer who looked for such tactics would be greatly disappointed. No attention is paid to the base of the support. It is not vouchsafed even a stroke of the rake. Nothing is done to overturn it, nothing, absolutely nothing! It is by other methods that the Burying-beetles obtain the Mole.

These decisive experiments, repeated under many different forms, prove that never, never in this world, do the Necrophori dig, or even give a superficial scrape, at the foot of the gallows, unless the hanging body touch the ground at that point. And, in the latter case, if the twig should happen to fall, this is in no way an intentional result, but a mere fortuitous effect of the burial already commenced.

What, then, did the man with the Frog, of whom Gleditsch tells us, really see? If his stick was overturned, the body placed to dry beyond the assaults of the Necrophori must certainly have touched the soil: a strange precaution against robbers and damp! We may well attribute more foresight to the preparer of dried Frogs and allow him to hang his animal a few inches off the ground. In that case, as all my experiments emphatically declare, the fall of the stake undermined by the sextons is a pure matter of imagination.

Yet another of the fine arguments in favour of the reasoning-power of insects flies from the light of investigation and founders in the slough of error! I wonder at your simple faith, O masters who take seriously the statements of chance-met observers, richer in imagination than in veracity; I wonder at your credulous zeal, when, without criticism, you build up your theories on such absurdities!

Let us continue. The stake is henceforth planted perpendicularly, but the body hanging on it does not reach the base: a condition enough to ensure that there will never be any digging at this point. I make use of a Mouse, who, by reason of her light weight, will lend herself better to the insect's manoeuvres. The dead animal is fixed by the hind-legs to the top of the apparatus with a raffia strap. It hangs plumb, touching the stick.

Soon two Necrophori have discovered the morsel. They climb the greased pole; they explore the prize, poking their foreheads into its fur. It is recognized as an excellent find. To work, therefore. Here we have again, but under more difficult conditions, the tactics employed when it was necessary to displace the unfavourably situated body: the two collaborators slip between the Mouse and the stake and, taking a grip of the twig and exerting a leverage with their backs, they jerk and shake the corpse, which sways, twirls about, swings away from the stake and swings back again. All the morning is passed in vain attempts, interrupted by explorations on the animal's body.

In the afternoon, the cause of the check is at last recognized; not very clearly, for the two obstinate gallow-robbers first attack the Mouse's hind-legs, a little way below the strap. They strip them bare, flay them and cut away the flesh about the foot. They have reached the bone, when one of them finds the string of raffia beneath his mandibles. This, to him, is a familiar thing, representing the grass-thread so frequent in burials in turfy soil. Tenaciously the shears gnaw at the bond; the fibrous fetter is broken; and the Mouse falls, to be buried soon after.

If it stood alone, this breaking of the suspending tie would be a magnificent performance; but considered in connection with the sum of the Beetle's customary labours it loses any far-reaching significance. Before attacking the strap, which was not concealed in any way, the insect exerted itself for a whole morning in shaking the body, its usual method. In the end, finding the cord, it broke it, as it would have broken a thread of couch-grass encountered underground.

Under the conditions devised for the Beetle, the use of the shears is the indispensable complement of the use of the shovel; and the modicum of discernment at his disposal is enough to inform him when it will be well to employ the clippers. He cuts what embarrasses him, with no more exercise of reason than he displays when lowering his dead Mouse underground. So little does he grasp the relation of cause and effect that he tries to break the bone of the leg before biting the raffia which is knotted close beside him. The difficult task is attempted before the extremely easy one.

Difficult, yes, but not impossible, provided that the Mouse be young. I begin over again with a strip of iron wire, on which the insect's shears cannot get a grip, and a tender Mousekin, half the size of an adult. This time a tibia is gnawed through, sawed in two by the Beetle's mandibles, near the spring of the heel. The detached leg leaves plenty of space for the other, which readily slips from the metal band; and the little corpse falls to the ground.

But, if the bone be too hard, if the prize suspended be a Mole, an adult Mouse or a Sparrow, the wire ligament opposes an insurmountable obstacle to the attempts of the Necrophori, who, for nearly a week, work at the hanging body, partly stripping it of fur or feather and dishevelling it until it forms a lamentable object, and at last abandon it when desiccation sets in. And yet a last resource remained, one as rational as infallible: to overthrow the stake. Of course, not one dreams of doing so.

For the last time let us change our artifices. The top of the gibbet consists of a little fork, with the prongs widely opened and measuring barely two-fifths of an inch in length. With a thread of hemp, less easily attacked than a strip of raffia, I bind the hind-legs of an adult Mouse together, a little above the heels; and I slip one of the prongs in between. To bring the thing down one has only to slide it a little way upwards; it is like a young Rabbit hanging in the window of a poulterer's shop.

Five Necrophori come to inspect what I have prepared. After much futile shaking, the tibiæ are attacked. This, it seems, is the method usually employed when the corpse is caught by one of its limbs in some narrow fork of a low-growing plant. While trying to saw through the bone—a heavy job this time—one of the workers slips between the shackled legs; in this position, he feels the furry touch of the Mouse against his chine. No more is needed to arouse his propensity to thrust with his back. With a few heaves of the lever the thing is done: the Mouse rises a little, slides over the supporting peg and falls to the ground.

Is this manoeuvre really thought out? Has the insect indeed perceived, by the light of a flash of reason, that to make the morsel fall it was necessary to unhook it by sliding it along the peg? Has it actually perceived the mechanism of the hanging? I know some persons—indeed, I know many—who, in the presence of this magnificent result, would be satisfied without further investigation.

More difficult to convince, I modify the experiment before drawing a conclusion. I suspect that the Necrophorus, without in any way foreseeing the consequences of his action, heaved his back merely because he felt the animal's legs above him. With the system of suspension adopted, the push of the back, employed in all cases of difficulty, was brought to bear first upon the point of support; and the fall resulted from this happy coincidence. That point, which has to be slipped along the peg in order to unhook the object, ought really to be placed at a short distance from the Mouse, so that the Necrophori may no longer feel her directly on their backs when they push.

A wire binds together now the claws of a Sparrow, now the heels of a Mouse and is bent, three-quarters of an inch farther away, into a little ring, which slips very loosely over one of the prongs of the fork, a short, almost horizontal prong. The least push of this ring is enough to bring the hanging body to the ground; and because it stands out it lends itself excellently to the insect's methods. In short, the arrangement is the same as just now, with this difference, that the point of support is at a short distance from the animal hung up.

The Glow-Worm and Other Beetles

My trick, simple though it be, is quite successful. For a long time the body is repeatedly shaken, but in vain; the tibiæ, the hard claws refuse to yield to the patient saw. Sparrows and Mice grow dry and shrivel, unused, upon the gallows. My Necrophori, some sooner, some later, abandon the insoluble mechanical problem: to push, ever so little, the movable support and so to unhook the coveted carcase.

Curious reasoners, in faith! If, just now, they had a lucid idea of the mutual relations between the tied legs and the suspending peg; if they made the Mouse fall by a reasoned manoeuvre, whence comes it that the present artifice, no less simple than the first, is to them an insurmountable obstacle? For days and days they work on the body, examining it from head to foot, without noticing the movable support, the cause of their mishap. In vain I prolong my watch; I never see a single one of them push the support with his foot or butt it with his head.

Their defeat is not due to lack of strength. Like the Geotrupes, they are vigorous excavators. When you grasp them firmly in your hand, they slip into the interstices of the fingers and plough up your skin so as to make you quickly loose your hold. With his head, a powerful ploughshare, the Beetle might very easily push the ring off its short support. He is not able to do so, because he does not think of it; he does not think of it, because he is devoid of the faculty attributed to him, in order to support their theories, by the dangerous generosity of the evolutionists.

Divine reason, sun of the intellect, what a clumsy slap in thy august countenance, when the glorifiers of the animal degrade thee with such denseness!

Let us now examine the mental obscurity of the Necrophori under another aspect. My captives are not so satisfied with their sumptuous lodging that they do not seek to escape, especially when there is a dearth of labour, that sovran consoler of the afflicted, man or beast. Internment within the wire cover palls upon them. So, when the Mole is buried and everything in order in the cellar, they stray uneasily over the trellised dome; they clamber up, come down, go up again and take to flight, a flight which instantly becomes a fall, owing to collision with the wire grating. They pick themselves up and begin all over again. The sky is splendid; the weather is hot, calm and propitious for those in search of the Lizard crushed beside the footpath. Perhaps the effluvia of the gamy tit-bit have reached them from afar, imperceptible to any other sense than that of the grave-diggers. My Necrophori therefore would be glad to get away.

Can they? Nothing would be easier, if a glimmer of reason were to aid them. Through the trelliswork, over which they have so often strayed, they have seen, outside, the free soil, the promised land which they want to reach. A hundred times if once have they dug at the foot of the rampart. There, in vertical wells, they take up their station, drowsing whole days on end while unemployed. If I give them a fresh Mole, they emerge from their retreat by the entrance-corridor and come to hide themselves beneath the belly of the beast. The burial over, they return, one here, one there, to the confines of the enclosure and disappear underground.

Well, in two and a half months of captivity, despite long stays at the base of the trellis, at a depth of three-quarters of an inch beneath the surface, it is rare indeed for a Necrophorus to succeed in circumventing the obstacle, in prolonging his excavation beneath the barrier, in digging an elbow and bringing it out on the other side, a trifling task for these vigorous creatures. Of fourteen only one succeeds in escaping.

A chance deliverance and not premeditated; for, if the happy event had been the result of a mental combination, the other prisoners, practically his equals in powers of perception, would all, from first to last, have discovered by rational means the elbowed path leading to the outer world; and the cage would promptly be deserted. The failure of the great majority proves that the single fugitive was simply digging at random. Circumstances favoured him; and that is all. We must not put it to his credit that he succeeded where all the others failed.

We must also beware of attributing to the Necrophori a duller understanding than is usual in insect psychology. I find the ineptness of the undertaker in all the Beetles reared under the wire cover, on the bed of sand into which the rim of the dome sinks a little way. With very rare exceptions, fortuitous accidents, not one thinks of circumventing the barrier by way of the base; not one manages to get outside by means of a slanting tunnel, not even though he be a miner by profession, as are the Dung-beetles *par excellence*. Captives under the wire dome and anxious to escape, Sacred Beetles, Geotrupes, Copres, Gymnopleuri,[3] Sisyphi,[4] all see about them the free space, the joys of the open sunlight; and not one thinks of going round under the rampart, which would present no difficulty to their pickaxes.

[3] Cf. *The Sacred Beetle and Others:* chap. vii.—*Translator's Note.*

[4] Cf. *idem:* chap. xv.—*Translator's Note.*

Even in the higher ranks of animality, examples of similar mental obfuscation are not lacking. Audubon[5] tells us how, in his days, wild Turkeys were caught in North America. In a clearing known to be frequented by these birds, a great cage was constructed with stakes driven into the ground. In the centre of the enclosure opened a short tunnel, which dipped under the palisade and returned to the surface outside the cage by a gentle slope, which was open to the sky. The central opening, wide enough to give a bird free passage, occupied only a portion of the enclosure, leaving around it, against the circle of stakes, a wide unbroken zone. A few handfuls of maize were scattered in the interior of the trap, as well as round about it, and in particular along the sloping path, which passed under a sort of bridge and led to the centre of the contrivance. In short, the Turkey-trap presented an ever-open door. The bird found it in order to enter, but did not think of looking for it in order to go out.

[5] John James Audubon (1780-1851), the noted American ornithologist, of French descent, author of *Birds of America* (1827-1830) and *Ornithological Biography* (1831-1839).—*Translator's Note.*

The Glow-Worm and Other Beetles

According to the famous American ornithologist, the Turkeys, lured by the grains of maize, descended the insidious slope, entered the short underground passage and beheld, at the end of it, plunder and the light. A few steps farther and the gluttons emerged, one by one, from beneath the bridge. They distributed themselves about the enclosure. The maize was abundant; and the Turkeys' crops grew swollen.

When all was gathered, the band wished to retreat, but not one of the prisoners paid any attention to the central hole by which he had arrived. Gobbling uneasily, they passed again and again across the bridge whose arch was yawning beside them; they circled round against the palisade, treading a hundred times in their own footprints; they thrust their necks, with their crimson wattles, through the bars; and there, with their beaks in the open air, they fought and struggled until they were exhausted.

Remember, O inept one, what happened but a little while ago; think of the tunnel that led you hither! If that poor brain of yours contains an atom of ability, put two ideas together and remind yourself that the passage by which you entered is there and open for your escape! You will do nothing of the kind. The light, an irresistible attraction, holds you subjugated against the palisade; and the shadow of the yawning pit, which has but lately permitted you to enter and will quite as readily permit you to go out, leaves you indifferent. To recognize the use of this opening you would have to reflect a little, to recall the past; but this tiny retrospective calculation is beyond your powers. So the trapper, returning a few days later, will find a rich booty, the entire flock imprisoned!

Of poor intellectual repute, does the Turkey deserve his name for stupidity? He does not appear to be more limited than another. Audubon depicts him as endowed with certain useful ruses, in particular when he has to baffle the attacks of his nocturnal enemy, the Virginian Owl. As for his behaviour in the snare with the underground passage, any other bird, impassioned of the light, would do the same.

Under rather more difficult conditions, the Necrophorus repeats the ineptness of the Turkey. When he wishes to return to the daylight, after resting in a short burrow against the rim of the cover, the Beetle, seeing a little light filtering through the loose soil, reascends the entrance-well, incapable of telling himself that he has only to prolong the tunnel as far in the opposite direction to reach the outer world beyond the wall and gain his freedom. Here again is one in whom we shall seek in vain for any sign of reflection. Like the rest, in spite of his legendary renown, he has no guide but the unconscious promptings of instinct.

CHAPTER XIII
THE GIANT SCARITES

The military profession can hardly be said to favour the talents. Consider the Carabus, or Ground-beetle, that fiery warrior among the insect people. What can he do? In the way of industry, nothing or next to nothing. Nevertheless the dull butcher is magnificent in his indescribably sumptuous jerkin. It has the refulgency of copper pyrites, of gold, of Florentine bronze. While clad in black, he enriches his sombre costume with a vivid amethyst hem. On the wing-cases, which fit him like a cuirass, he wears little chains of alternate pins and bosses.

Of a handsome and commanding figure, slender and pinched in at the waist, the Carabus is the glory of our collections, but only for the sake of his appearance. He is a frenzied murderer; and that is all. We will ask nothing more of him. The wisdom of antiquity represented Hercules, the god of strength, with the head of an idiot. And indeed merit is not great when limited to brute force. And this is the case with the Carabus.

To see him so richly adorned, who would not wish to find him a fine subject for investigation, one worthy of history, a subject such as humbler natures provide with lavish generosity? From this ferocious ransacker of entrails we expect nothing of the kind. His art is that of slaying.

We may without trouble observe him at his bandit's work. I rear him in a large breeding-cage on a layer of fresh sand. A few potsherds scattered about the surface enable him to take shelter beneath the rocks; a tuft of grass planted in the centre makes a grove and enlivens the establishment.

Three species compose the population: the common *Jardinière*, or Golden Beetle, the usual inmate of our gardens; *Procrustes coriaceus*, the sombre and powerful explorer of the grassy thickets at the foot of walls; and the rare Purple Carabus, who trims the ebony of his wing-cases with metallic violet. I feed them on Snails, after partly removing the shell.

Hidden at first promiscuously under the potsherds, the Carabi make a rush for the wretched Snail, who, in his despair, alternately puts out and withdraws his horns. Three of them at a time, then four, then five begin by devouring the edge of his mantle, specked with chalky atoms. This is the favourite morsel. With their mandibles, those stout pincers, they lay hold of it through the froth; they tug at it, tear off a shred and retire to a distance to swallow it at their ease.

Meanwhile the legs, streaming with slime, pick up grains of sand and become covered with heavy gaiters, which are extremely cumbersome but to which the Beetle pays no attention. Heavy with mire, he staggers back to his prey and cuts off another morsel. He will think of polishing his boots presently. Others do not stir, but gorge themselves on the spot, with the whole fore-part of their body immersed in the froth. The feast lasts for hours on end. The guests do not leave the joint until the distended belly lifts the roof of the wing-cases and uncovers the nudities of the stern.

Fonder of shady nooks, the Procrustes form a separate company. They drag the Snail into their lair, under the shelter of a potsherd, and there, peacefully and in common, dismember the mollusc. They love the Slug, as easier to cut up than the Snail, who is defended by his shell; they regard the Testacella,[1] who bears a chalky shell, shaped like a Phrygian cap, right at the hinder end of her foot, as a delicious tit-bit. The game has firmer flesh and is less nauseously slimy.

[1] Or Shell-bearing Slug, found along the shores of the Mediterranean.—*Translator's Note.*

To feast gluttonously on a Snail whom I myself have rendered defenseless by breaking her shell is nothing for a warrior to boast about; but we shall soon see the Carabus display his daring. I offer a Pine-chafer, in the pink of strength, to the Golden Beetle, whose appetite has been whetted by a few days' fasting. The victim is a colossus beside the Golden Carabus; an Ox facing a Wolf.

The beast of prey prowls round the peaceful creature and selects its moment. It rushes forward, recoils, hesitates and returns to the charge. And lo, the giant is overthrown! Incontinently the other devours him, ransacking his belly. If this had happened in a higher order of the animal world, it would make one's flesh creep to watch the Carabus half immersed in the big Cockchafer and rooting out his entrails.

I test the eviscerator with a more difficult quarry. This time the victim is *Oryctes nasicornis*, the powerful Rhinoceros Beetle, an invincible giant, one would think, under the shelter of his armour. But the hunter knows the weak point of the horn-clad prey, the fine skin protected by the wing-cases. By means of attacks which the assailant renews as soon as they are repulsed by the assailed, the Carabus contrives to raise the cuirass slightly and to slip his head beneath it. From the moment that the pincers have made a gash in the vulnerable skin, the Rhinoceros is lost. Soon there will be nothing left of the colossus but a pitiful empty carcase.

Those who wish for a more hideous conflict must apply to *Calosoma sycophanta*, the handsomest of our flesh-eating insects, the most majestic in costume and size. This prince of Carabi is the butcher of the caterpillars. He is not to be overawed even by the sturdiest of rumps.

His struggle with the huge caterpillar of the Great Peacock Moth[2] is a thing to see once, not oftener: a single experience of such horrors is enough to disgust one. The contortions of the eviscerated insect, which, with a sudden heave of the loins, hurls the bandit in the air and lets him fall, belly uppermost, without managing to make him release his hold; the green entrails spilt quivering on the ground; the tramping gait of the murderer, drunk with slaughter, slaking his thirst at the springs of a horrible wound: these are the main features of the combat. If entomology had no other scenes to show us, I should without the least regret turn my back upon my insects.

[2] Cf. *The Life of the Caterpillar:* chap. xi.—*Translator's Note.*

Next day, offer the sated Beetle a Green Grasshopper or a White-faced Decticus, serious adversaries both, armed with powerful lower jaws. With these big-bellied creatures the slaughter will begin anew, as eagerly as on the day before. It will be repeated later with the Pine-chafer and the Rhinoceros Beetle, accompanied by the usual atrocious tactics of the Carabi. Even better than these last does the Calosoma know the weak point of the armoured Beetles, concealed beneath the wing-cases. And this will go on so long as we keep him provided with victims, for this drinker of blood is never satiated.

Acrid exhalations, the products of a fiery temperament, accompany this frenzy for carnage. The Carabi elaborate caustic humours; the Procrustes squirts a jet of vinegar at any one who takes hold of him; the Calosoma makes the fingers smell of mouldy drugs; certain Beetles, such as the Brachini,[3] understand explosives and singe the aggressor's whiskers with a volley of musketry.

[3] Or Bombardier Beetles. When disturbed, they eject a fluid which volatilizes, on contact with the air, with a slight report.—*Translator's Note.*

Distillers of corrosives, gunners throwing lyddite, bombers employing dynamite: what can all these violent creatures, so well equipped for battle, do beyond committing slaughter? Nothing. We find no art, no industry, not even in the larva, which practices the adult's trade and meditates its crimes while wandering under the stones. Nevertheless it is to one of these dull-witted warriors that I am deliberately proposing to apply to-day, prompted by the wish to solve a certain question. Let me tell you what it is.

You have surprised this or that insect, motionless on a bough, blissfully basking in the sun. Your hand is raised, open, ready to descend on it and seize it. Hardly have you made the movement when the insect drops to the ground. It is a wearer of armoured wing-cases, slow to disengage the wings from their horny sheath, or perhaps an incomplete form, with no wing-surfaces. Incapable of sudden flight, the surprised insect lets itself fall. You look for it in the grass, often in vain. If you do find it, it is lying on its back, with its legs folded, without stirring.

It is shamming dead, people will tell you; it is pretending, in order to escape its enemy. Man is certainly unknown to it; we count for nothing in its little world. What does it care for our hunting, whether we be children or scientists? It does not fear the collector with his long pin; but it realizes danger in general; and it dreads its natural enemy, the insectivorous bird, which swallows it with a single snap. To outwit the assailant, it lies upon its back, draws up its legs and simulates death. The bird, or any other persecutor, will despise it in this condition; and its life will be saved.

This, we are assured, is how the insect would reason if suddenly surprised. The trick has long been famous. Once upon a time, two friends, at the end of their resources, sold the skin of a Bear before they had killed the brute. The encounter was unfortunate: they had to take to their heels. One of them stumbled, fell, held his breath and shammed dead. The Bear came up, turned the man over and over, explored him with his paw and his muzzle, sniffed at his face:

"He smells already," he said and, without more ado, turned away.

That Bear was a simpleton.

The bird would not be duped by this clumsy stratagem. In those happy days when the discovery of a nest marked a red-letter day, I never saw my Sparrows or Greenfinches refuse a Locust because he was not moving, or a Fly because she was dead. Any mouthful that does not kick is eagerly accepted, provided that it be fresh and pleasant to the taste.

If the insect, therefore, relies on the appearance of death, it would seem to me to be very badly inspired. More wary than the Bear in the fable, the bird, with its perspicacious eye, will recognize the fraud in a moment and proceed to business. Besides, had the object really been a corpse, but still fresh, it would none the less have gobbled it up.

More insistent doubts occur to my mind when I consider the serious consequences to which the insect's artfulness might lead. It shams dead, says the popular idiom, which recks little of weighing the value of its term; it simulates death, scientific language repeats, happy to find some gleams of reason in the insect. What truth is there in this unanimous statement, which in the one case is too unreflecting and in the other too much inclined to favour theoretical fancies?

Logical arguments are insufficient here. It is essential that we should obtain the verdict of experiment, which alone can furnish a valid reply. But to which of the insects shall we go first?

I remember something that dates back some forty years. Delighted with a recent University triumph, I was staying at Cette, on my return from Toulouse, where I had just passed my examination as a licentiate in natural science. It gave me a fine chance of renewing my acquaintance with the seaside flora, which had delighted me a few years before on the shores of the wonderful Gulf of Ajaccio. It would have been foolish to neglect it. A degree does not confer the right to cease studying. If one really has a touch of the sacred fire in one's veins, one remains a student all one's life, not of books, which are a poor resource, but of the great, inexhaustible school of actual things.

One day, then, in July, in the cool stillness of the dawn, I was botanizing on the foreshore at Cette. For the first time I plucked the *Convolvulus soldanella*, which trails along the high-water mark its ropes of glossy green leaves and its great pink bellflowers. Withdrawn into his white, flat, heavily-keeled shell, a curious Snail, *Helix explanata*, was slumbering, in groups, on the bent grasses.

The dry shifting sands showed here and there long series of imprints, recalling, on a smaller scale and under another form, the tracks of little birds in the snow which used to arouse a delightful flutter in my youthful days. What do these imprints mean?

I follow them, a hunter on the trail of a new species. At the end of each track, by digging to no great depth, I unearth a magnificent Carabus, whose very name is almost unknown to me. It is the Giant Scarites (*S. gigas*, FAB.).

I make him walk on the sand. He exactly reproduces the tracks which put me on the alert. It was certainly he who, questing for game in the night, marked the trail with his feet. He returned to his lair before daylight; and now not a single Beetle is to be seen in the open.

Another characteristic thrusts itself upon my notice. If I shake him for a moment and then place him on the ground upon his back, he remains a long time without stirring. No other insect has yet displayed such persistent immobility, though I confess that my investigations in this respect have been only superficial. The detail is so thoroughly engraved on my memory that, forty years later, when I want to experiment on the insects which are experts in the art of simulating death, I at once think of the Scarites.

A friend sends me a dozen from Cette, from the very beach on which I once passed a delightful morning in the company of this skilful mimic of the dead. They reach me in perfect condition, mixed up in the same package with some Pimeliæ (*P. bipunctata*, FAB.), their compatriots in the sands beside the sea. Of these last, a pitiable crew, many have been disembowelled, absolutely emptied; others have merely stumps instead of legs; a few, but only a few, are unwounded.

It was what one might have expected of these Carabidæ, lawless hunters one and all. Tragic events took place in the box during the journey from Cette to Sérignan. The Scarites gormandized riotously on the peaceable Pimeliæ.

Their tracks, which I followed long ago on the actual spot, bore evidence to their nocturnal rounds, apparently in search of their prey, the pot-bellied Pimelia, whose sole defence consists of a strong armour of welded wing-cases.[4] But what can such a cuirass avail against the bandit's ruthless pincers?

[4] The Pimelia is a wingless Beetle. — *Translator's Note.*

He is indeed a mighty hunter, this Nimrod of the sea-shore. All black and glossy, like a jet bugle, his body is divided by a very narrow groove at the waist. His weapon of offence consists of a pair of claw-like mandibles of extraordinary vigour. None of our insects equals him in strength of jaw, if we except the Stag-beetle, who is far better armed, or rather decorated, for the antlered mandibles of the inmate of the oak are ornaments of the male's attire, not a panoply of battle.

The brutal Carabid, the eviscerator of the Pimeliæ, knows how strong he is. If I tease him a little on the table, he at once adopts a posture of defence. Well braced upon his short legs, especially the fore-legs, which are toothed like rakes, he dislocates himself in two, so to speak, thanks to the groove that divides him behind the corselet; he proudly raises the fore-part of the body, his wide, heart-shaped thorax and massive head, opening his threatening pincers to their full extent. He is now an awesome sight. More: he has the audacity to rush at the finger which has touched him. Here of a surety is one not easily intimidated. I look twice before I handle him.

I lodge my strangers partly under a wire-gauze cover and partly in glass jars, all supplied with a layer of sand. Each of them without delay digs himself a burrow. The insect bends his head a long way down and, with the points of his mandibles, brought together to form a pick-axe, he hews, digs and excavates with a will. The fore-legs, spread out and armed with hooks, gather the dust and rubbish into a load which is thrust backwards. In this way, a mound rises on the threshold of the burrow. The dwelling grows deeper quickly and by a gentle slope reaches the bottom of the jar.

Checked in the downward direction, the Scarites now digs against the glass wall and continues his work horizontally until he has obtained a length of nearly twelve inches in all.

This arrangement of the gallery, almost the whole of which runs just under the glass, is very useful to me, enabling me to follow the insect in the privacy of its home. If I wish to observe its underground operations, all that I need do is to remove the opaque sheath which I have been careful to put over the jar, in order to spare the creature the annoyance of the light.

When the house is deemed to be long enough, the Scarites returns to the entrance, which he works more carefully than the rest. He makes a funnel of it, a pit with shifting, sloping sides. It is the Ant-lion's crater on a larger scale and constructed in a more rustic fashion. This mouth is continued by an inclined plane, kept free of all rubbish. At the foot of the slope is the vestibule of the horizontal gallery. Here, as a rule, the hunter lurks, motionless, with his pincers half open. He is waiting.

There is a sound overhead. It is a specimen of game which I have just introduced, a Cicada, a luscious morsel. The drowsy trapper at once wakes; he moves his palpi, which quiver with cupidity. Cautiously, step by step, he climbs his inclined plane. He takes a glance outside the funnel. The Cicada is seen.

The Scarites darts out of his pit, runs forward, seizes the Cicada and drags her backwards. The struggle is brief, thanks to the trap of the entrance, which yawns like a funnel to receive even a bulky quarry and contracts into a crumbling precipice that paralyses all resistance. The slope is fatal: who crosses the brink can no longer escape the murderer.

Head first, the Cicada dives into the abyss, down which the spoiler drags her by successive jerks. She is drawn into the low-ceilinged tunnel. Here the wings cease to flutter, for lack of space. She reaches the knacker's cellar, at the end of the corridor. The Scarites now works at her for some time with his pincers, in order to reduce her to complete immobility, fearing lest she should escape; then he returns to the mouth of the charnel-house.

It is not everything to possess plenty of game; the question next arises how to consume it in peace. The door is therefore closed against importunate callers, that is to say, the insect fills the entrance to the tunnel with his mound of rubbish. Having taken this precaution, he goes back again and sits down to his meal. He will not reopen his hiding-place nor remake the pit at the entrance until later, when the Cicada has been digested and hunger makes its reappearance. Let us leave the glutton with his quarry.

The brief morning which I spent with him in his native place did not enable me to watch him at his hunting, on the sands of the beach; but the facts gathered in captivity are enough to tell us all about it. They show us in the Scarites a bold hero who is not to be intimidated by the biggest or strongest adversary.

We have seen him coming up from underground, falling on the passers-by, seizing them at some distance from the burrow and dragging them forcibly into his cut-throat den. The Rose-chafer, the Common Cockchafer are but small deer for him. He dares to attack the Cicada, he dares to dig his hooks into the corpulent Pine-chafer. He is a fearless ruffian, ready for any crime.

Under natural conditions his audacity can be no less. On the contrary, the familiar spots, freedom of movement, unlimited space and his beloved salt air excite the warrior to yet greater feats of daring.

He has dug himself a refuge in the sand, with a wide, crumbling mouth. This is not so that he may, like the Ant-lion, wait at the bottom of his funnel for the passing of a victim which stumbles on the shifting slope and rolls into the pit. The Scarites disdains these petty poachers' methods, these fowlers' snares; he prefers a run across country.

His long trails on the sand tell us of nocturnal rounds in search of big game, often the Pimelia, sometimes the Half-spotted Scarab.[5] The find is not consumed on the spot. To enjoy it at his ease, he needs the peaceful darkness of the underground manor; and so the captive, seized by one leg with the pincers, is forcibly dragged along the ground.

[5] Cf. *The Sacred Beetle and Others:* chaps. ii. and vii. — *Translator's Note.*

If no precautions were taken, the introduction of the victim into the burrow would be impracticable, with a huge quarry offering a desperate resistance. But the entrance to the tunnel is a wide crater, with crumbling walls. However large he be, the captive, tugged from below, enters and tumbles into the pit. The crumbling rubbish immediately buries him and paralyses his movements. The thing is done. The bandit now proceeds to close his door and empty his prey's belly.

CHAPTER XIV
THE SIMULATION OF DEATH

The first insect that we will put to the question is that audacious disemboweller, the savage Scarites. To provoke his state of inertia is a very simple matter: I handle him for a moment, rolling him between my fingers; better still, I drop him on the table, twice or thrice in succession, from a small height. When the shock due to the fall has been administered and, if need be, repeated, I turn the insect on its back.

This is enough: the prostrate Beetle no longer stirs, lies as though dead. The legs are folded on the belly, the antennæ extended like the arms of a cross, the pincers open. A watch beside me tells me the exact minute of the beginning and the end of the experiment. Nothing remains but to wait and especially to arm one's self with patience, for the insect's immobility lasts long enough to become tedious to the observer watching for something to happen.

The duration of the lifeless posture varies greatly on the same day, under the same atmospheric conditions and with the same subject, though I cannot fathom the causes which shorten or lengthen it. How to investigate the external influences, so numerous and often so slight, which intervene in such a case; above all, how to scrutinize the insect's private impressions: these are impenetrable mysteries. Let us confine ourselves to recording the results.

Immobility continues fairly often for as long as fifty minutes; in certain cases, even, it lasts more than an hour. The most frequent length of time averages twenty minutes. If nothing disturbs the Beetle, if I cover him with a glass shade, protecting him from the Flies, who are importunate visitors in the hot weather prevailing at the time of my experiment, the inertia is complete: not a quiver of the tarsi, nor of the palpi, nor of the antennæ. Here indeed is a simulacrum of death, with all its inertia.

At last the apparently deceased comes back to life. The tarsi quiver, those of the fore-legs first; the palpi and the antennæ move slowly to and fro: this is the prelude to the awakening. Now the legs begin to kick. The insect bends slightly at its pinched waist; it buttresses itself on its head and back; it turns over. There it goes, jogging away, ready to become an apparent corpse once more if I renew my shock tactics.

Let us repeat the experiment immediately. The newly resuscitated Beetle is for a second time lying motionless on his back. He prolongs his make-believe of death longer than he did at first. When he wakes up, I renew the test a third, a fourth, a fifth time, with no intervals of repose. The duration of the motionless condition increases each time. To quote the figures, the five consecutive experiments, from the first to the last, have continued respectively for 17, 20, 25, 33 and 50 minutes. Starting with a quarter of an hour, the attitude of death ends by lasting nearly a whole hour.

Without being constant, similar facts recur repeatedly in my experiments, the duration, of course, varying. They tell us that as a general rule the Scarites lengthens the period of his lifeless posture the oftener the experiment is repeated. Is this a matter of practice, or is it an increase of cunning employed in the hope of finally tiring a too persistent enemy? It would be premature to draw conclusions: the cross-examination of the insect has not yet been thorough enough.

Let us wait. Besides, we need not imagine that it is possible to go on like this until our patience is exhausted. Sooner or later, flurried by my pestering, the Scarites refuses to sham dead. Scarcely is he laid on his back after a fall, when he turns over and takes to his heels, as though he judged a stratagem which succeeded so indifferently to be henceforth useless.

If we were to stop here, it would certainly seem that the insect, a cunning hoaxer, seeks, as a means of defence, to cheat those who attack him. He counterfeits death; he repeats the process, becoming more persistent in his fraud in proportion as the aggression is repeated; he abandons his trickery when he deems it futile. But hitherto we have subjected him only to a friendly examination-in-chief. The time has come to put a string of searching questions and to trick the trickster if there be really any deception.

The Beetle under experiment is lying on the table. He feels beneath him a hard body which gives him no chance of digging. As he cannot hope to take refuge underground, an easy task for his nimble and vigorous tools, the Scarites lies low in his death-like pose, keeping it up, if need be, for an hour. If he were reclining on the sand, the loose soil with which he is so familiar, would he not regain his activity more rapidly, would he not at least betray by a few twitches his desire to escape into the basement?

I was expecting to see him do so; and I was mistaken. Whether I place him on wood, glass, sand or garden mould, the Beetle in no way modifies his tactics. On a surface readily excavated he continues his immobility as long as on an unassailable surface.

This indifference to the nature of the support half opens the door to doubt; what follows opens it wide. The patient is on the table before me and I watch him closely. With his gleaming eyes, overshadowed by his antennæ, he also sees me; he watches me; he observes me, if I may so express myself. What can be the visual impression of the insect when face to face with that monstrosity, man? How does the pigmy measure the enormous monument that is the human body? Seen from the depths of the infinitely little, the immense perhaps is nothing.

We will not go so far as that; we will admit that the insect watches me, recognizes me as his persecutor. So long as I am here, he will suspect me and refuse to budge. If he does decide to do so, it will be after he has exhausted my patience. Let us therefore move away. Then, since any trickery will be needless, he will hasten to take to his legs again and make off.

I move ten paces farther from him, to the other end of the room. I hide, I do not move a muscle, for fear of breaking the silence. Will the insect pick itself up? No, my precautions are superfluous. Alone, left to itself, perfectly quiet, it remains motionless for as long a time as when I was standing close beside it.

Perhaps the clear-sighted Scarites has seen me in my corner, at the other end of the room; perhaps a subtle scent has revealed my presence to him. We will do more, then. I cover him with a bell-glass which will save him from being worried by the Flies and I leave the room; I go downstairs into the garden. There is no longer anything likely to disturb him. Doors and windows are closed. Not a sound from without; no cause for alarm indoors. What will happen in the midst of that profound silence?

Nothing more and nothing less than usual. After twenty, forty minutes' waiting out of doors, I come upstairs again and return to my insect. I find him as I left him, lying motionless on his back.

This experiment, many times repeated with different subjects, throws a vivid light upon the question. It expressly assures us that the attitude of death is not the ruse of an insect in danger. Here there is nothing to alarm the creature. Around him all is silence, solitude, repose. When he persists in his immobility it cannot now be to deceive an enemy. I have no doubt about it: there is something else involved.

Besides, why should he need special defensive artifices? I could understand that a weak, pacific, ill-protected insect might resort to ruses when in danger; but in him, the warlike bandit, so well armoured, it is more than I can understand. No insect on his native sea-shore has the strength to resist him. The most powerful of them, the Sacred Beetle and the Pimelia, are easy-going creatures which, so far from molesting him, are fine booty for his burrow.

Can he be threatened by the birds? It is very doubtful. As a Carabus, he is saturated with acrid humours which must make his body a far from pleasing mouthful. For the rest, he lives hidden from the light of day in a burrow where no one sees him; he emerges only at night, when the birds are no longer inspecting the beach. There are no beaks about for him to fear.

And this butcher of the Pimeliæ and even occasionally of the Sacred Beetles, this bully whom no danger threatens, is supposed to be such a coward as to sham death on the slightest alarm! I take the liberty of doubting this more and more.

I am confirmed in my doubts by the Smooth-skinned Scarites (*S. lavigatus*, FAB.), a denizen of the same shores. The first insect is a giant; the second, by comparison, is a dwarf. Otherwise he displays the same shape, the same jet-black costume, the same armour, the same habits of brigandage. Well, the Smooth-skinned Scarites, in spite of his weakness and his smallness, is almost ignorant of the trick of pretending to be dead. When molested for a moment and then turned on his back, he at once picks himself up and flees. I can hardly obtain a few seconds' immobility; once only, daunted by my obstinacy, the dwarf remains motionless for a quarter of an hour.

How different from the giant, motionless the moment that he is thrown upon his back, sometimes picking himself up only after an hour of inaction! It is the reverse of what ought to happen, if the apparent death were really a defensive ruse. The giant, confident in his strength, should disdain this cowardly posture; the timid dwarf should be quick to have recourse to it. And it is just the other way about. What is there behind all this?

Let us try the influence of danger. With what natural enemy shall I confront the big Scarites, motionless on his back? I know none. Let us then create a make-believe assailant. The Flies put me on the track of one.

I have spoken of their importunity during my investigations in the hot season. If I do not employ a bell-glass or keep an assiduous watch, rarely does the shrewish Dipteron fail to alight upon my patient and explore him with her proboscis. We will let her have her way this time.

Hardly has the Fly grazed this apparent corpse with her legs, when the Scarites' tarsi quiver as though twitched by a slight electric shock. If the visitor be merely passing, matters go no farther; but, if she persist, particularly near the Beetle's mouth, moist with saliva and disgorged secretions of food, the tormented Scarites promptly kicks, turns over and makes off.

Perhaps he did not think it opportune to prolong his fraud in the face of so contemptible an enemy. He resumes his activity because he has recognized the absence of danger. Then let us call in another interloper, one of formidable size and strength. I happen to have handy a Great Capricorn, with powerful claws and mandibles. That the long-horned insect is a peaceful creature I am well aware; but the Scarites does not know it; on the sands of the shore he has never encountered such a colossus as this, who is capable of impressing less timid creatures than he. Fear of the unknown will merely aggravate the situation.

Guided by the tip of my straw, the Capricorn sets his foot upon the prostrate insect. The Scarites' tarsi begin to quiver immediately. If the contact be prolonged or multiplied, or if it become aggressive, the dead insect gets on its legs again and scuttles off, just as the titillations of the Fly have already shown me. When danger is imminent and all the more to be dreaded because its nature is unknown, the trick of the simulation of death disappears and flight takes its place.

The following experiment is not without value. I take some hard substance and knock the foot of the table on which the insect is lying on its back. The shock is very slight, not enough to shake the table perceptibly. The whole thing is limited to the inner vibrations of a resilient body which has received a blow. But it is quite enough to disturb the insect's immobility. At each tap the tarsi are flexed and quiver for a moment.

Lastly, let us try the effect of light. So far, the patient has been treated in the shade of my cabinet, away from the direct sunlight. The sun is shining full upon the window. What will the motionless insect do if I carry it thither, from my table to the window, into the bright light? That we can find out in a moment. Under the direct rays of the sun, the Scarites immediately turns over and moves off.

This is enough. Patient, persecuted creature, you have half-betrayed your insect. When the Fly tickles you, drains your moist lip, treats you as a corpse whose juices she would like to suck; when the huge Capricorn appears to your horrified gaze and puts a foot on your belly, as though to take possession of his prey; when the table quivers, that is to say, when, for you, the ground shakes, undermined perhaps by some invader of your burrow; when a bright light surrounds you, favouring the designs of your enemies and imperilling your safety as an insect that loves the dark, then, in truth, it would be wiser not to move, if really your chief resource, when danger threatens you, is to simulate death.

On the contrary, at those critical moments, you give a start; you move, you resume your normal attitude, you run away. Your fraud is discovered; or, to put it more plainly, there is no trick. Your inertia is not simulated; it is real. It is a condition of temporary torpor into which you are plunged by your delicate nervous organization. A mere nothing makes you fall into it; a mere nothing withdraws you from it, above all a bath of light, that sovran stimulus of activity.

In respect of prolonged immobility as the result of emotion, I find a rival of the Giant Scarites in a large black Buprestis, with a flour-speckled corselet, a lover of the blackthorn, the hawthorn and the apricot-tree. His name is *Capnodis tenebrionis*, LIN. At times I see him, with his legs closely folded and his antennæ lowered, prolonging his motionless posture upon his back for more than an hour. At other times the insect is bent upon escaping, apparently influenced by atmospheric conditions of which I do not know the secret. One or two minutes' immobility is as much as I can then obtain.

Let me recapitulate: in my various subjects the attitude of death is of very variable duration, governed as it is by a host of unsuspected circumstances. Let us take advantage of favourable opportunities, which are fairly frequent. I subject the Cloudy Buprestis to the different tests undergone by the Giant Scarites. The results are the same. When you have seen the first, you have seen the second. There is no need to linger over them.

I will only mention the promptness with which the Buprestis, lying motionless in the shade, recovers his activity when I carry him away from my table into the broad sunlight of the window. After a few seconds of this bath of heat and light, the insect half-opens his wing-cases, using them as levers, and turns over, ready to take flight if my hand did not instantly snap him up. He is a passionate lover of the light, a devotee of the sun, intoxicating himself in its rays upon the bark of his blackthorn-trees on the hottest afternoons.

This love of tropical temperature suggests the following question: what would happen if I were to chill the creature in its immobile posture? I foresee a more prolonged inertia. The chill, of course, must not be great, for it would be followed by the lethargy into which insects capable of surviving the winter fall when benumbed by the cold.

On the contrary, the Buprestis must as far as possible retain his full vitality. The lowering of the temperature must be gentle, very moderate and such that the insect, under similar climatic conditions, would retain his powers of action in ordinary life. I have a convenient refrigerator at my disposal. It is the water of my well, whose temperature, in summer, is nearly twenty-two degrees Fahrenheit below that of the surrounding air.

The Buprestis, in whom I have just produced inertia by means of a few taps, is installed on his back in a little flask which I seal hermetically and immerse in a bucket full of this cold water. To keep the bath as cool as at first, I gradually renew it, taking care not to shake the flask in which the patient is lying, in his attitude of death.

The result rewards my pains. After five hours under water, the insect is still motionless. Five hours, I say, five long hours; and I might certainly say longer, if my exhausted patience had not put an end to the experiment. But this is enough to banish any idea of fraud on the insect's part. Here, beyond a doubt, the insect is not shamming dead. He is actually somnolent, deprived of the power of movement by an internal disturbance which my teasing produced at the outset and which is prolonged beyond its usual limits by the surrounding coolness.

I try the effect of a slight decrease in temperature upon the Giant Scarites by subjecting him to a similar sojourn in the cold water of the well. The result does not respond to the hopes which the Buprestis gave me. I do not succeed in obtaining more than fifty minutes' inertia. I have often obtained as long periods of immobility without resorting to the refrigerating artifice.

It might have been foreseen. The Buprestis, a lover of the burning sunshine, is affected by the cold bath in a different degree from the Scarites, who prowls about by night and spends his day in the basement. A fall of a few degrees in temperature takes the chilly insect by surprise and has no effect upon the one accustomed to the coolness underground.

Other experiments on these lines tell me nothing more. I see the inert condition persisting sometimes for a longer, sometimes for a shorter period, according as the insect seeks the sunlight or avoids it. Let us change our method.

I evaporate a few drops of sulphuric ether in a glass jar and put in a Stercoraceous Geotrupes and a specimen of *Buprestis tenebrionis*, at the same time. In a few moments both subjects are motionless, anæsthetized by the etheric vapour. I take them out quickly and lay them on their backs in the open air.

Their attitude is exactly that which they would have assumed under the influence of a shock or any other cause of alarm. The Buprestis has his legs symmetrically folded against his chest and belly; the Geotrupes has his outspread, stretched in disorder, rigid and as though attacked by catalepsy. You could not tell if they were dead or alive.

They are not dead. In a minute or two, the Geotrupes' tarsi twitch, the palpi quiver, the antennæ wave gently to and fro. Then the forelegs move; and a quarter of an hour has not elapsed before the other legs are struggling. The activity of the insect made motionless by the concussion of a shock would reawaken in precisely the same fashion.

As for the Buprestis, he is in a state of inertia so profound that at first I really believe him to be dead. He recovers during the night; and next day I find him in possession of his usual activity. The ether experiment, which I took care to stop at the moment when it produced the desired effect, has not been fatal to him; but it has had much more serious consequences for him than for the Geotrupes. The insect more sensitive to the alarm due to concussion or to a fall of temperature is also the more sensitive to the action of ether.

Thus the enormous difference which I observe in these two insects, with regard to the inertia provoked by a shock or by handling them in one's fingers, is explained by nice differences of impressionability. Whereas the Buprestis remains motionless for nearly an hour, the Geotrupes is struggling violently after a minute or two. And even then I rarely attain this limit.

In what respect has the Geotrupes, to defend itself, less need of the stratagem of simulated death than the Black Buprestis, well protected by his massive build and his armour, which is so hard that it resists the point of a pin and even of a needle? We should be perplexed by the same question in respect of a multitude of insects, some of which remain motionless while others do not; and we could not possibly foresee what would happen from the genus of the subject, its form, or its way of living.

Buprestis tenebrionis, for example, exhibits a persistent inertia. Will it be the same, because of similarity of structure, with other members of the same group? Not at all. My chance finds provide me with the Brilliant Buprestis (*B. rutilans*, FAB.), and the Nine-spotted Buprestis (*Ptosima novemmaculata*, FAB.). The first resists all my attempts. The splendid creature grips my fingers, grips my tweezers and insists on getting up the moment that I lay it on its back. The second readily becomes immobile; but how brief is its attitude of death! Four or five minutes at most.

A Melasoma-beetle, *Omocrates abbreviatus*, OLIV., whom I frequently discover under the broken stones on the neighbouring hills, continues motionless for over an hour. He rivals the Scarites. We must not forget to add that very often the awakening takes place within a few minutes.

Can he owe his long period of inertia to the fact that he is one of the Tenebrionidæ, or Darkling Beetles? By no means, for here in the same group is *Pimelia bipunctata*, who turns a somersault on his round back and finds his feet the moment he has turned over; here is a Cellar-beetle (*Blaps similis*, LATR.), who, unable to turn with his flat back, his big belly and his welded wing-cases,[1] struggles desperately after a minute or two of inertia.

[1] The Cellar-beetle is one of the wingless Beetles.—*Translator's Note.*

The short-legged Beetles, trotting along with tiny steps, ought, one would think, to make up in cunning, more fully than the others, for their incapacity for rapid flight. The facts do not correspond with this apparently well-founded forecast. I have consulted the genera Chrysomela,[2] Blatta,[3] Silpha, Cleonus,[4] Bolboceras,[5] Cetonia, Hoplia, Coccinella,[6] and so on. A few minutes or a few seconds are nearly always long enough for the return to activity. Several of them even obstinately refuse to sham death.

[2] Golden-apple Beetles.—*Translator's Note.*

[3] Blackbeetles or Cockroaches.—*Translator's Note.*

[4] A genus of Weevils.—*Translator's Note.*

[5] A mushroom-eating Beetle. Cf. *The Life of the Fly:* chap. xviii.—*Translator's Note.*

[6] Ladybirds.—*Translator's Note.*

As much must be said of the Beetles well-equipped for pedestrian escape. Some remain motionless for a few seconds; others, more numerous still, behave in an ungovernable fashion. In short, there is no guide to tell us in advance:

"This one will readily assume the posture of a dead insect; this one will hesitate; that one will refuse."

There is nothing but shadowy probabilities, until experiment has given its verdict. From this muddle shall we draw a conclusion which will set our minds at rest? I hope so.

CHAPTER XV
SUICIDE OR HYPNOSIS?

You do not imitate the unfamiliar; you do not counterfeit a thing of which you know nothing: that is obvious. The simulation of death, therefore, implies a certain knowledge of death.

Well, has the insect, or rather, has any kind of animal, a presentiment that its life cannot last for ever? Does the perturbing problem of an end occur to its dense brain? I have associated a great deal with animals, I have lived on intimate terms with them and I have never observed anything to justify me in saying yes. The animal, with its humbler destiny, is spared that apprehension of the hour of death which constitutes at once our torment and our greatness.

Like the child still in the limbo of unconsciousness, it enjoys the present without taking thought of the future; free from the bitterness of a prospective ending, it lives in the blissful calm of ignorance. It is ours alone to foresee the briefness of our days; it is ours alone anxiously to question the grave regarding the last sleep.

Moreover, this glimpse of the inevitable destruction calls for a certain maturity of mind and, for that reason, is rather late in developing. I had a touching example of it this very week.

A pretty little Kitten, the joy of all the household, after languidly dragging itself about for a couple of days, died in the night. Next morning the children found it lying stark in its basket. General affliction. Anna, especially, a little girl of four, considered with a pensive glance the little friend with which she had so often played. She petted it, called it, offered it a drop of milk in a cup:

"Kitty won't play," said the child. "She doesn't want my breakfast any more. She's asleep. I've never seen her sleep like this before. When will she wake up?"

This simplicity in the presence of death's harsh problem wrung my heart. Hastily I led the girl away from the sight and had the dead Kitten secretly buried. As, from this time onward, it no longer appeared by the table at meal-times, the grief-stricken child at last understood that she had seen her little friend sleeping the profound slumber that knows no awaking. For the first time a vague idea of death found its way into her mind.

Has the insect the signal honour of knowing what we do not know in our early childhood, at a time when thought is already manifesting itself, far superior, however feeble it be, to the dull understanding of the animal? Has it the power to foresee an ending, an attribute which in its case would be inconvenient and useless? Before deciding, let us consult, not the abstruse theories of science, a doubtful guide, but the Turkey, an eminently truthful one.

I recall one of the most vivid memories that remain to me from my brief sojourn at the Royal College of Rodez. So they called it then; to-day they call it a grammar-school; what improvement as the world grows older!

The thrice-blessed Thursday had come; our bit of translation was done, our dozen Greek roots had been learnt by heart; and we trooped down to the far end of the valley, so many bands of madcaps. With our trousers turned up to our knees, we exploited, artless fishermen that we were, the peaceful waters of the river, the Aveyron. What we hoped to catch was the Loach, no bigger than our little finger, but tempting, thanks to his immobility on the sand amid the waterweeds. We fully expected to transfix him with our trident, a fork.

This miraculous catch, the object of such shouts of triumph when it succeeded, was very rarely vouchsafed to us: the Loach, the rascal, saw the fork coming and with three strokes of his tail disappeared!

We found compensation in the apple-trees in the neighbouring pastures. The apple has from all time been the urchin's delight, above all when plucked from a tree which does not belong to him. Our pockets were soon crammed with the forbidden fruit.

Another distraction awaited us. Flocks of Turkeys were not rare, roaming at their own sweet will and gobbling up the Locusts around the farms. If no watcher hove in sight, we had great sport. Each of us would seize a Turkey, tuck her head under her wing, rock it in this attitude for a moment and then place her on the ground, lying on her side. The bird no longer budged. The whole flock of Turkeys was subjected to our hypnotic handling; and the meadow assumed the aspect of a battle-field strewn with the dead and dying.

And now look out for the farmer's wife! The loud gobbling of the harassed birds had told her of our wicked pranks. She would run up armed with a whip. But we had good legs in those days! And we had a good laugh too, behind the hedges, which favoured our retreat!

O delightful days when we put the Turkeys to sleep, can I recover the skill which I then possessed? To-day it is no longer the playful trick of a schoolboy; it is a matter of serious research. I happen to have the very subject that I need: a Turkey-hen, doomed soon to be the victim of our Christmas merry-making. I repeat with her the method of manipulation which I employed so successfully on the banks of the Aveyron. I tuck her head well under her wing and, molding it in this attitude with both hands, I rock the bird gently up and down for a couple of minutes.

The strange effect is produced; my childhood's manoeuvres obtained no better result. Laid on the ground, on her side and left to herself, my patient is a lifeless bundle. One would think her dead, if a slight rise and fall of the plumage did not reveal the breathing. She looks really like a dead bird which, in a last convulsion, had drawn its chilled feet, with their shrivelled toes, under its belly. The spectacle has a tragic air; and I feel overcome by a certain anxiety when I gaze upon the results of my evil spells. Poor Turkey! What if she were never to wake again!

We need not be afraid: she is waking; she stands up, staggering a little, it is true, with drooping tail and a shamefaced expression. That soon passes off; not a trace of it remains. In a few moments the bird is once more what it was before the experiment.

This torpor, the mean between true sleep and death, is of variable duration. When repeatedly provoked in my Turkey-hen, with suitable intervals of repose, immobility lasts sometimes for half an hour and sometimes for a few minutes. Here, as in the insect, it would be very difficult to analyse the causes of these differences. With the Guinea-fowl I succeed even better. The torpor lasts so long that I become alarmed by the bird's condition. The plumage reveals no trace of breathing. I ask myself, anxiously, whether the bird is not actually dead. I push it a little way along the ground with my foot. The patient does not stir. I do it again. And lo, the Guinea-fowl frees her head, stands up, regains her balance and scurries off! Her state of lethargy has lasted more than half an hour.

Now for the Goose. I have none. The gardener next door trusts me with his. She is brought to my house, which she fills with her trumpeting as she waddles about. Shortly afterwards there is absolute silence: the web-footed Amazon is lying on the ground, with her head tucked under her wing. Her immobility is as profound and as prolonged as that of the Turkey and the Guinea-fowl.

It is the Hen's turn now and the Duck's. They too succumb, but, so it seems to me, less persistently. Can it be that my hypnotic tricks are less efficacious with small birds than with large ones? To judge by the Pigeon, this may well be so. He yields to my art only to the extent of two minutes' sleep. A still smaller bird, a Greenfinch, is even more refractory: all that I obtain from him is a few seconds' drowsiness.

It would appear, then, that, in proportion as the activity is concentrated in a body of less volume, the torpor has less hold. The insect has already shown us this. The Giant Scarites does not stir for an hour, while the Smooth-skinned Scarites, a pigmy, wearies my persistence in turning him over; the large Cloudy Buprestis submits to my manoeuvres for a long period, whereas the Glittering Buprestis, a pigmy again, obstinately refuses to do so.

We will leave on one side, as insufficiently investigated, the influence of the bodily mass and remember only this fact, that it is possible, by a very simple artifice, to reduce a bird to a condition of apparent death. Do my Goose, my Turkey and the others resort to trickery with the object of deceiving their tormentor? It is certain that none of them thinks of shamming dead; they are actually immersed in a deep torpor; in a word, they are hypnotized.

These facts have long been known; they are perhaps the first in date in the science of hypnosis or artificial sleep. How did we, the little Rodez schoolboys, learn the secret of the Turkey's slumber? It was certainly not in our books. Coming from no one knows where, indestructible as everything that enters into children's games, it was handed down, from time immemorial, from one initiate to another.

Things are just the same to-day in my village of Sérignan, where there are numbers of youthful adepts in the art of putting poultry to sleep. Science often has very humble beginnings. There is nothing to tell us that the mischief of a pack of idle urchins is not the starting-point of our knowledge of hypnosis.

I have just been practising on insects tricks which to all appearances are as puerile as those which we practised on the Turkeys in the days when the farmer's wife used to run after us cracking her whip. Do not laugh: a serious problem looms behind this artlessness.

My insects' condition bears a strange resemblance to that of my poultry. Both present the image of death, inertia, the contraction of convulsed limbs. In both again the immobility is dispelled before its time by the agency of a stimulus, by sound in the case of the bird, by light in that of the insect. Silence, darkness and tranquillity prolong it. Its duration varies greatly in different species and appears to increase with corpulence.

Among ourselves, who are very unequal subjects for induced sleep, the hypnotist is obliged to pick and choose. He succeeds with one and not with another. Similarly, among the insects, a selection is necessary, for they do not all of them, by a long way, respond to the experimenter's attempts. My best subjects have been the Giant Scarites and the Cloudy Buprestis; but how many others have resisted quite indomitably, or remained motionless for only a few seconds!

The insect's return to the active state presents certain peculiarities which are well worthy of attention. The key to the problem lies here. Let us return for a moment to the patients who have been subjected to the ordeal of ether. These are really hypnotized. They do not remain motionless by way of a ruse, there is no doubt upon that point; they are actually on the threshold of death; and, if I did not take them in good time out of the flask in which a few drops of ether have been evaporated, they would never recover from the torpor whose last stage is death.

Now what symptoms herald their return to activity? We know the symptoms: the tarsi tremble, the palpi quiver, the antennæ wave to and fro. A man emerging from a deep sleep stretches his limbs, yawns and rubs his eyes. The insect awaking from the etheric sleep likewise has its own fashion of marking its recovery of consciousness: it flutters its tiny digits and the more mobile of its organs.

Let us now consider an insect which, upset by a shock, perturbed by some sort of excitement, is believed to be shamming dead, lying on its back. The return to activity is announced exactly in the same fashion and in the same order as after the stupefying effect of ether. First the tarsi quiver; then the palpi and antennæ wave feebly to and fro.

If the creature were really shamming, what need would it have of these minute preliminaries to the awakening? Once the danger has disappeared, or is deemed to have done so, why does the insect not swiftly get upon its feet, to make off as quickly as possible, instead of dallying with untimely pretences? I am quite sure that, once the Bear was gone, the comrade who had shammed dead under the animal's nose did not think of wasting time in stretching himself or rubbing his eyes. He jumped up at once and took to his heels.

And the insect is supposed to carry its cunning to the length of counterfeiting resuscitation down to the least details! No, no and again no; it would be madness. Those quiverings of the tarsi, those awakening movements of the palpi and antennæ are the obvious proof of a genuine torpor, now coming to an end, a torpor similar to that induced by ether but less intense; they show that the insect struck motionless by my artifice is not shamming dead, as the vulgar idiom has it and as the fashionable theories repeat. It is really hypnotized.

A shock which disturbs its nerve-centres, an abrupt fright which seizes upon it reduce it to a state of somnolence like that of the bird which is swung for a second or two with its head under its wing. A sudden terror sometimes deprives us human beings of the power of movement, sometimes kills us. Why should not the insect's organism, so delicate and subtle, give way beneath the grip of fear and momentarily succumb? If the emotion be slight, the insect shrinks into itself for an instant, quickly recovers and makes off; if it be profound, hypnosis supervenes, with its prolonged immobility.

The insect, which knows nothing of death and therefore cannot counterfeit it, knows nothing either of suicide, that desperate means of cutting short excessive misery. No authentic example has ever been given, to my knowledge, of an animal of any kind robbing itself of its own life. That those most richly endowed with the capacity of affection sometimes allow themselves to die of grief I grant you; but there is a great difference between this and stabbing one's self or cutting one's throat.

Yet the recollection occurs to me of the Scorpion's suicide, sworn to by some, denied by others. What truth is there in the story of the Scorpion who, surrounded by a circle of fire, puts an end to his suffering by stabbing himself with his poisoned sting? Let us see for ourselves:

Circumstances favour me. I am at this moment rearing, in large earthen pans, with a bed of sand and with potsherds for shelter, a hideous menagerie which hardly comes up to my expectations as regards the study of morals.[1] I will profit by it in another way. It consists of some twenty-four specimens of *Buthus occitanus*, the large White Scorpion of the south of France. The odious animal abounds, always isolated, under the flat stones of the neighbouring hills, in the sandy spots which enjoy the most sunlight. It has a detestable reputation.

[1] For the habits of the White or Languedocian Scorpion, cf. *The Life and Love of the Insect:* chaps. xvii. and xviii.—*Translator's Note.*

On the effects of its sting I personally have nothing to say, having always avoided, by a little caution, the danger to which my relations with the formidable captives in my study might have exposed me. Knowing nothing of it myself, I get people to tell me of it, woodcutters in particular, who from time to time fall victims to their imprudence. One of them tells me the following story:

"After having my dinner, I was dozing for a moment among my faggots, when I was roused by a sharp pain. It was like the prick of a red-hot needle. I clapped my hand to the place. Sure enough, there was something moving! A Scorpion had crept under my trousers and stung me in the lower part of the calf. The ugly beast was full as long as my finger. Like that, sir, like that!"

And, adding gesture to speech, the worthy man extended his great fore-finger. This size did not surprise me: while insect-hunting, I have seen Scorpions as large.

"I wanted to go on with my work," he continued, "but I came out in a cold sweat; and my leg swelled up so you could see it swelling. It got as big as that, sir, as big as that."

More mimicry. Our friend spreads his two hands round his leg, at a distance, so as to denote the girth of a small barrel:

"Yes, like that, sir, like that; I had great trouble to get home, though it was only half a mile away. The swelling crept up and up. Next day it had got so high."

A gesture indicates the height.

"Yes, sir, for three days I couldn't stand up. I bore it as well as I could, with my leg stretched out on a chair. Soda-compresses did the trick; and there you are, sir, there you are."

Another woodcutter, he adds, was also stung in the lower part of the leg. He was binding faggots together at some distance and had not the strength to regain his home. He collapsed by the side of the road. Some men passing by carried him on their shoulders:

"*À la cabro morto, moussu, à la cabro morto!*"

The story of the rustic narrator, more versed in mimicry than in speech, does not seem to me exaggerated. A White Scorpion's sting is a very serious accident for a human being. When stung by his own kind, the Scorpion himself quickly succumbs. Here I have something better than the evidence of strangers: I have my own observations.

I take two healthy specimens from my menagerie and place them together at the bottom of a glass jar on a layer of sand. Excited with the tip of a straw which brings them face to face again whenever they draw back, the two harassed creatures decide on mortal combat. Each no doubt attributes to the other the annoyances of which I myself am the cause. The claws, those weapons of defence, are displayed in a semicircle and open to keep the adversary at a distance; the tails, in sudden jerks, are flung forward above the back; the poison-phials clash together; a tiny drop, limpid as water, beads the point of the sting.

The fight does not last long. One of the Scorpions receives the full force of the other's poisoned weapon. It is all over: in a few minutes the wounded one succumbs. The victor very calmly proceeds to gnaw the fore-part of the victim's cephalothorax, or, in less crabbed terms, the bit at which we look for a head and find only the entrance to a belly. The mouthfuls are small, but long-drawn-out. For four or five days, almost without a break, the cannibal nibbles at his murdered comrade. To eat the vanquished, that's good warfare, the only sort excusable. What I do not understand, nor shall until we tin the meat on the battle-field for food, is our wars between nations.

We now have authentic information: the Scorpion's sting is fatal, promptly fatal, to the Scorpion himself. Let us come to the matter of suicide, such as it has been described to us. When surrounded by a circle of live embers, the animal, so we are told, stabs itself with its sting and finds an end of its torment in voluntary death. This would be very fine on the creature's part if it were true. We shall see.

In the centre of a ring of burning charcoal, I place the largest specimen from my menagerie. The bellows increase the glow. At the first smart of the heat, the animal moves backwards within the circle of fire. It collides by inadvertence with the burning barrier. Now follows a disorderly retreat, in every direction, at random, renewing the agonizing contact. At each attempt to escape, the burning is repeated more severely than before. The animal becomes frantic. It darts forward and scorches itself. In a desperate frenzy, it brandishes its weapon, crooks it, straightens it, lays it down flat and raises it again, all with such disorderly haste that I am quite unable to follow its movements accurately.

The moment ought to have come for the Scorpion to release himself from his torture with a blow of the stiletto. And indeed, with a sudden spasm, the long-suffering creature becomes motionless, lies at full-length, flat upon the ground. There is not a movement; the inertia is complete. Is the Scorpion dead? It really looks like it. Perhaps he has pinked himself with a thrust of his sting that escaped me in the turmoil of the last efforts. If he has actually stabbed

The Glow-Worm and Other Beetles

himself, if he has resorted to suicide, then he is dead beyond a doubt: we have just seen how quickly he succumbs to his own venom.

In my uncertainty, I pick up the apparently dead body with the tip of my forceps and lay it on a bed of cool sand. An hour later, the alleged corpse returns to life, as lusty as before the ordeal. I repeat the process with a second and third specimen. The results are the same. After the frantic plunges of the desperate victim, we have the same sudden inertia, with the creature sprawling flat as though struck by lightning, and the same return to life on the cool sand.

It seems probable that those who invented the story of the Scorpion committing suicide were deceived by this sudden swoon, this paralysing spasm, into which the high temperature of the enclosure throws the exasperated beast. Too quickly convinced, they left the victim to burn to death. Had they been less credulous and withdrawn the animal in good time from its circle of fire, they would have seen the apparently dead Scorpion return to life and thus assert its profound ignorance of suicide.

Apart from man, no living thing knows the last resource of a voluntary end, because none has a knowledge of death. As for us, to feel that we have the power to escape from the miseries of life is a noble prerogative, upon which it is good to meditate, as a sign of our elevation above the commonalty of the animal world; but in point of fact it becomes cowardice if from the possibility we pass to action.

He who proposes to go to that length should at least repeat to himself what Confucius, the great philosopher of the yellow race, said five-and-twenty centuries ago. Having surprised a stranger in the woods fixing to the branch of a tree a rope wherewith to hang himself, the Chinese sage addressed him in words the gist of which was as follows:

"However great your misfortunes, the greatest of all would be to yield to despair. All the rest can be repaired; this one is irreparable. Do not believe that all is lost for you and try to convince yourself of a truth which has been proved indisputable by the experience of the centuries. And that truth is this: so long as a man has life, there is no need for him to despair. He may pass from the greatest misery to the greatest joy, from the greatest misfortune to the highest felicity. Take courage and, as though you were this very day beginning to recognize the value of life, strive at every moment to make the most of it."

This humdrum Chinese philosophy is not without merit. It suggests the moralizing of the fabulist:

> "... Qu'on me rende impotent,
> Cul-de-jatte, goutteux, manchot, pourvu qu'en somme
> Je vive, c'est assez: je suis plus que content."[2]

> [2] "... So powerless let me lie,
> Gout-ridden, legless, armless; if only, after all,
> I live, it is enough: more than content am I."

Yes, yes, La Fontaine and Kung the philosopher are right: life is a serious matter, which it will not do to throw away into the first bush by the roadside like a useless garment. We must look upon it not as a pleasure, nor yet as a punishment, but as a duty of which we have to acquit ourselves as well as we can until we are given leave to depart.

To anticipate this leave is cowardly and foolish. The power to disappear at will through death's trap-door does not justify us in deserting our post; but it opens to us certain vistas which are absolutely unknown to the animal.

We alone know how life's pageant closes, we alone can foresee our end, we alone profess devotion to the dead. Of these high matters none other has any suspicion. When would-be scientists proclaim aloud, when they declare that a wretched insect knows the trick of simulating death, we will ask them to look more closely and not to confound the hypnosis due to terror with the pretence of a condition unknown to the animal world.

Ours alone is the clear vision of an end, ours alone the glorious instinct of the beyond. Here, filling its modest part, speaks the voice of entomology, saying:

"Have confidence; never did an instinct fail to keep its promises."

CHAPTER XVI
THE CRIOCERES

I am a stubborn disciple of St. Thomas the Apostle and, before I agree to anything, I want to see and touch it, not once, but twice, thrice, an indefinite number of times, until my incredulity bows beneath the weight of evidence. Well, the Rhynchites[1] have told us that the build does not determine the instincts, that the tools do not decide the trade. And now, yes, the Crioceres come and add their testimony. I question three of them, all common, too common, in my paddock. At the proper season, I have them before my eyes, without searching for them, whenever I want to ask them for information.

[1] A genus of Weevils, the essays upon whom will appear in a later volume to be entitled *The Life of Weevil.—Translator's Note.*

The first is the Crioceris of the Lily, or Lily-beetle. Since Latin words offend our modesty let us just once mention her scientific name, *Crioceris merdigera*, LIN., without translating it, or, above all, repeating it. Decency forbids. I have never been able to understand why natural history need inflict upon a lovely flower or an engaging animal an odious name.

As a matter of fact, our Crioceris, so ill-treated by the nomenclators, is a sumptuous creature. She is nicely shaped, neither too large nor too small, and a beautiful coral red, with jet-black head and legs. Everybody knows her who in the spring has ever glanced at the lily, when its stem is beginning to show in the centre of the rosette of leaves. A Beetle, of less than the average size and coloured sealing-wax red, is perched up on the plant. Your hand goes out to seize her. Forthwith, paralysed with fright, she drops to the ground.

Let us wait a few days and return to the lily, which is gradually growing taller and beginning to show its buds, gathered together in a bundle. The red insect is still there. Further, the leaves, which are seriously bitten into, are reduced to tatters and soiled with little heaps of greenish ordure. It looks as if some witchcraft had mashed up the leaves and then splashed the mess all over the place.

Well, this filth moves, travels slowly along. Let us overcome our repugnance and poke the heaps with a straw. We uncover, indeed we unclothe an ugly, pot-bellied, pale-orange larva. It is the grub of the Crioceris.

The origin of the garment of which we have just stripped it would be unmentionable, save in the world of the insect, that manufacturer devoid of shame. This doublet is, in fact, obtained from the creature's excretions. Instead of evacuating downwards, on the superannuated principle, the Crioceris' larva evacuates upwards and receives upon its back the waste products of the intestine, materials which move from back to front as each fresh pat is dabbed upon the others. Réaumur has complacently described how the quilt moves forward from the tail to the head by wriggling along inclined planes, making so many dips in the undulating back. There is no need to return to this stercoral mechanism after the master has done with it.

We now know the reasons that procured the Lily-beetle an ignominious title, confined to the official records: the grub makes itself an overcoat of its excrements.

Once the garment is completed so as to cover the whole of the creature's dorsal surface, the clothing-factory does not cease work on that score. At the back a fresh hem is added from moment to moment; but the overlapping superfluity in front drops off of its own weight at the same time. The coat of dung is under continual repair, being renovated and lengthened at one end as it wears and grows shorter at the other.

Sometimes also the stuff is too thick and the heap capsizes. The denuded grub recks nothing of the lost overcoat; its obliging intestine repairs the disaster without delay.

Whether by reason of the clipping that results from the excessive length of a piece which is always on the loom, or of accidents that cause a part or the whole of the load to fall off, the grub of the Crioceris leaves accumulations of dirt in its track, till the lily, the symbol of purity, becomes a very cess-pool. When the leaves have been browsed, the stem next loses its cuticle, thanks to the nibbling of the grub, and is reduced to a ragged distaff. The flowers even, which have opened by now, are not spared: their beautiful ivory chalices are changed into latrines.

The perpetrator of the misdeed embarks on his career of defilement early. I wanted to see him start, to watch him lay the first course of his excremental masonry. Does he serve an apprenticeship? Does he work badly at first, then a little better and then well? I now know all about it: there is no noviciate, there are no clumsy attempts; the workmanship is perfect from the outset, the product ejected spreads over the hinder part. Let me tell you what I saw.

The eggs are laid in May, on the under surface of the leaves, in short trails averaging from three to six. They are cylindrical, rounded at both ends, of a bright orange-red, glossy and varnished with a glutinous wash which makes them stick to the leaves throughout their length. The hatching takes ten days. The shell of the egg, now a little wrinkled, but still of a bright orange colour, retains its position, so that the group of eggs, apart from its slightly withered appearance, remains just as it was.

The young larva measures a millimetre and a half[2] in length. The head and legs are black, the rest of the body a dull amber-red. On the first segment of the thorax is a brown sash, interrupted in the middle; lastly, there is a small black speck on each side, behind the third segment. This is the initial costume. Presently orange-red will take the place of the pale amber. The tiny creature, which is exceedingly fat, sticks to the leaf with its short legs and also with its hind-quarters, which act as a lever and push the round belly forwards. The motion reminds you of a cripple sitting in a bowl.

[2] .959 inch. — *Translator's Note.*

The grubs emerging from any one group of eggs at once begin to browse, each beside the empty skin of its egg. Here, singly, they nibble and dig themselves a little pit in the thickness of the leaf, while sparing the cuticle of the opposite surface. This leaves a translucent floor, a support which enables them to consume the walls of the excavation without risking a fall.

Seeking for better pasture, they move lazily on. I see them scattered at random; a few of them are grouped in the same trench; but I never see them browsing economically abreast as Réaumur relates. There is no order, no understanding between messmates, contemporaries though they be and all sprung from the same row of eggs. Nor is any heed paid to economy: the lily is so generous!

Meanwhile, the paunch swells and the intestine labours. Here we are! I see the first bit of the overcoat evacuated. As is natural in extreme infancy, it is liquid and there is not much of it. The scanty flow is used all the same and is laid methodically, right at the far end of the back. Let the little grub be. In less than a day, piece by piece, it will have made itself a suit.

The artist is a master from the first attempt. If its baby-flannel is so good to start with, what will the future ulster be, when the stuff, brought to perfection, is of much better quality? Let us proceed; we know what we want to know concerning the talents of this manufacturer of excremental broadcloth.

What is the purpose of this nasty great-coat? Does the grub employ it to keep itself cool, to protect itself against the attacks of the sun? It is possible: a tender skin need not be afraid of blistering under such a soothing poultice. Is it the grub's object to disgust its enemies? This again is possible: who would venture to set tooth to such a heap of filth? Or can it be simply a caprice of fashion, an outlandish fancy? I will not say no. We have had the crinoline, that senseless bulwark of steel hoops; we still have the extravagant stove-pipe hat, which tries to mould our heads in its stiff sheath. Let us be indulgent to the evacuator nor disparage his eccentric wardrobe. We have eccentricities of our own.

To feel our way a little in this delicate question, we will question the near kinsmen of the Lily-beetle. In my acre or two of pebbles I have planted a bed of asparagus. The crop, from the culinary point of view, will never repay me for my trouble: I am rewarded in another fashion. On the scanty shoots which I allow to display themselves freely in plumes of delicate green, two Crioceres abound in the spring: the field species (*C. campestris*, LIN.) and the twelve-spotted species (*C. duodecimpunctata*, LIN.). A splendid windfall, far better than any bundle of asparagus.

The first has a tricolor costume which is not without merit. Blue wing-cases, braided with white on the outer edge and each adorned with three white dots; a red corselet, with a blue disk in the centre. Its eggs are olive-green and cylindrical and, instead of lying flat, grouped in short lines, after the manner of the lily-dweller's, occur singly and stand on end on the leaves of the asparagus-plant, on the twigs, on the flower-buds, more or less everywhere, without any fixed order.

Though living in the open air on the leaves of its plant and thus exposed to all the various perils that may threaten the Lily-grub, the larva of the Field Crioceris knows nothing whatever of the art of sheltering itself beneath a layer of ordure. It goes through life naked and always perfectly clean.

It is of a bright greenish yellow, fairly fat behind and thinner in front. Its principal organ of locomotion is the end of the intestine, which protrudes, curves like a flexible finger, clasps the twig and supports the creature while pushing it forward. The true legs, which are short and placed too far in front with regard to the length of the body, would find it very difficult by themselves to drag the heavy mass that comes after. Their assistant, the anal finger, is remarkably strong. With no support, the larva turns over, head downwards, and remains suspended when shifting from one sprig to another. This Jack-in-the-bowl is a rope-dancer, a consummate acrobat, performing its evolutions amid the slender sprigs without fear of a fall.

Its attitude in repose is curious. The heavy stern rests on the two hind-legs and especially on the crooked finger, the end of the intestine. The fore-part is lifted in a graceful curve, the little black head is raised and the creature looks rather like the crouching Sphinx of antiquity. This pose is common at times of slumber and blissful digestion in the sun.

An easy prey is this naked, plump, defenceless grub, snoozing in the heat of a blazing day. Various Gnats, of humble size, but very likely terribly treacherous, haunt the foliage of the asparagus. The larva of the Crioceris, motionless in its sphinx-like attitude, does not appear to be on its guard against them, even when they come buzzing above its rump. Can they be as harmless as their peaceful frolics seem to proclaim? It is extremely doubtful: the Fly rabble are not there merely to imbibe the scanty exudations of the plant. Experts in mischief, they have no doubt hastened hither with another object.

And, in truth, on the greater number of the Crioceris-larvæ we find, adhering firmly to the skin, certain white specks, very small and of a china-white. Can these be the sowing of a bandit, the spawn of a Midge?

I collect the grubs marked with these white specks and rear them in captivity. A month later, about the middle of June, they shrivel, wrinkle and turn brown. All that is left of them is a dry skin which tears from end to end, half uncovering a Fly-pupa. A few days later, the parasite emerges.

It is a small, greyish Fly, fiercely bristling with sparse hairs, half the size of the House-fly, whom it resembles slightly. It belongs to the Tachina group, who, in their larval form, so often inhabit the bodies of caterpillars.

The white spots sprinkled over the larva of the Crioceris were the eggs of the hateful Fly. The vermin born of those eggs have perforated the victim's paunch. By subtle wounds, which cause little pain and are almost immediately healed, they have penetrated the body, reaching the humours in which the entrails are bathed. At first the larva invaded is not aware of its danger; it continues to perform its rope-dancer's gymnastics, to fill its belly and to take its siestas in the sun, as though nothing serious had occurred.

Reared in a glass tube and often examined under the lens, my parasite-ridden larvæ betray no uneasiness. The fact is that the Tachina's children display an infernal judgment in their first actions. Until the moment when they are ready for the transformation, their portion of game has to hold out, must be kept fresh and alive. They therefore gorge themselves with the reserves intended for future use, the fats, the savings which the Crioceris hoards in view of the remodelling whence the perfect insect will emerge; they consume what is not essential to the life of the moment and are very careful not to touch the organs which are indispensable at the present time. If these received a bite, the host would die and so would they. Towards the end of their growth, prudence and discretion being no longer essential, they make a complete clearance of the victim, leaving only the skin, which will serve them for a shelter.

One satisfaction is vouchsafed me in these horrible orgies: I see that the Tachina in her turn is subjected to severe reductions. How many were there on the larva's back? Perhaps eight, ten or more. One Midge, never more than one, comes out of the victim's skin, for the morsel is too small to provide food for many. What has become of the others? Has there been an internecine battle inside the poor wretch's body? Have they eaten one another up, leaving only the strongest to survive, or the one most favoured by the chances of the fight? Or has one of them, earlier developed than the rest, found himself master of the stronghold and have the others preferred to die outside rather than enter a grub already occupied, where famine would be rife if the messmates numbered even two? I am all for mutual extermination. Kinsman's flesh or stranger's flesh must be all one to the fangs of the vermin swarming in the Crioceris' belly.

Fierce though the competition is among these bandits, the Beetle's race does not threaten to die out. I review the innumerable troop on my asparagus-bed. A good half of them have Tachina-eggs plainly visible as tiny white specks on their green skins. The blemished larvæ tell me of a paunch already or on the point of being invaded. On the other hand, it is doubtful whether those which are unscathed will all remain in that condition. The malefactor is incessantly prowling around the green plumes, watching for a favourable opportunity. Many larvæ free from white spots to-day will show them to-morrow or some other day, so long as the Fly's season lasts.

I estimate that the vast majority of the troop will end by being infested. My rearing-experiments tell me much on this point. If I do not make a careful selection when I am stocking my wire-gauze-covers, if I go to work at random in picking the branches colonized with larvæ, I obtain very few adult Crioceres; nearly all of them are resolved into a cloud of Midges.

If it were possible for us to wage war effectually upon an insect, I should advise asparagus-growers to have recourse to the Tachina, though I should cherish no illusions touching the results of the expedient. The exclusive tastes of the insect auxiliary draw us into a vicious circle: the remedy allays the evil, but the evil is inseparable from the remedy. To rid ourselves of the ravages of the asparagus-beds, we should need a great many Tachinæ; and to obtain a great many Tachinæ we should first of all need a great many ravagers. Nature's equilibrium balances things as a whole. Whenever Crioceres abound, the Midges that reduce them arrive in numbers; when Crioceres become rare, the Midges decrease, but are always ready to return in masses and repress a surplus of the others during a return of prosperity.

Under its thick mantle of ordure the grub of the Lily-beetle escapes the troubles so fatal to its cousin of the asparagus. Strip it of its overcoat: you will never find the terrible white specks upon its skin. The method of preservation is most effective.

Would it not be possible to find a defensive system of equal value without resorting to detestable filth? Yes, of course: the insect need only house itself under a covering where there would be nothing to fear from the Fly's eggs. This is what the Twelve-spotted Crioceris does, occupying the same quarters as the Field Crioceris, from whom she differs in size, being rather larger, and still more in her costume, which is rusty red all over, with twelve black spots distributed symmetrically on the wing-cases.

Her eggs, which are a deep olive-green and cylindrical, pointed at one pole and squared off at the other, closely resemble those of the Field Crioceris and, like these, usually stand up on the supporting surface, to which they are fastened by the square end. It would be easy to confuse the two if we had not the position which they occupy to guide us. The Field Crioceris fastens her eggs to the leaves and the thin sprays; the other plants them exclusively on the still green fruit of the asparagus, globules the size of a pea.

The grubs have to open a tiny passage for themselves and to make their own way into the fruit, of which they eat the pulp. Each globule harbours one larva, no more, or the ration would be insufficient. Often, however, I see two, three or four eggs on the same fruit. The first grub hatched is the one favoured by luck. He becomes the owner of the pill, an intolerant owner capable of wringing the neck of any who should come and sit down at table beside him. Always and everywhere this pitiless competition!

The grub of the Twelve-spotted Crioceris is a dull white, with an interrupted black scarf on the first segment of the thorax. This sedentary creature has none of the talents of the acrobat grazing on the swaying foliage of the asparagus; it cannot take a grip with its posterior, turned into a prehensile finger. What use would it have for such a prerogative, loving repose as it does and destined to put on fat in its cell, without roaming in quest of food? In the same group each species has its own gifts, according to the kind of life that awaits it.

It is not long before the occupied fruit falls to the ground. Day by day, it loses its green colour as the pulp is consumed. It becomes, at last, a pretty, diaphanous opal sphere, while the berries which have not been injured ripen on the plant and acquire a rich scarlet hue.

When there is nothing left to eat inside the skin of its pill, the grub makes a hole in it and goes underground. The Tachinæ have spared it. Its opal box, the hard rind of the berry, has ensured its safety just as well as a filthy overcoat would have done and perhaps even better.

CHAPTER XVII
THE CRIOCERES (*continued*)

The Crioceris has found safety inside its opal globe. Safety? Ah, but what an unfortunate expression I have used! Is there any one in the world who can flatter himself that he has escaped the spoiler?

In the middle of July, at the time when the Twelve-spotted Crioceris comes up from under the ground in the adult form, my rearing-jars yield me swarms of a very small Gall-fly, a slender, graceful, blue-black Chalcid, without any visible boring-tool. Has the puny creature a name? Have the nomenclators catalogued it? I do not know, nor do I much care; the main thing is to learn that the covering of the asparagus-berry, which becomes an opal globe when the grub has emptied it, has failed to save the recluse. The Tachina-midge drains her victim by herself; this other, tinier creature feasts in company. Twenty or more of them batten on the grub together.

When everything seems to foretell a quiet life, a pigmy among pigmies appears, charged with the express duty of exterminating an insect which is protected first by the casket of the berry and next by the shell, the underground work of the grub. To eat the Twelve-spotted Crioceris is its mission in life, its special function. When and how does it deliver its attack? I do not know.

At any rate, proud of her vocation and finding life sweet, the Chalcid curls her antennæ into a crook and waves them to and fro: she rubs her tarsi together, a sign of satisfaction; she dusts her belly. I can hardly see her with the naked eye; and yet she is an agent of the universal extermination, a wheel in the implacable machine which crushes life as in a wine-press.

The tyranny of the belly turns the world into a robber's cave. Eating means killing. Distilled in the alembic of the stomach, the life destroyed by slaughter becomes so much fresh life. Everything is melted down again, everything has a fresh beginning in death's insatiable furnace.

Man, from the alimentary point of view, is the chief brigand, consuming everything that lives or might live. Here is a mouthful of bread, the sacred food. It represents a certain number of grains of wheat which asked only to sprout, to turn green in the sun, to shoot up into tall stalks crowned with ears. They died that we might live. Here are some eggs. Left undisturbed with the Hen, they would have emitted the Chickens' gentle cheep. They died that we might live. Here is beef, mutton, poultry. Horror, it smells of blood, it is eloquent of murder! If we gave it a thought, we should not dare to sit down to table, that altar of cruel sacrifices.

How many lives does the Swallow, to mention only the most peaceable, harvest in the course of a single day! From morning to evening he gulps down Crane-flies, Gnats and Midges joyously dancing in the sunbeams. Quick as lightning he passes; and the dancers are decimated. They perish; then their melancholy remnants fall from the nest containing the young brood, in the form of guano which becomes the turf's inheritance. And so it is with all and everything, with large and small, from end to end of the animal progression. A perpetual massacre perpetuates the flux of life.

Appalled by these butcheries, the thinker begins to dream of a state of affairs which would free us from the horrors of the maw. This ideal of innocence, as our poor nature vaguely sees it, is not an impossibility; it is partly realized for all of us, men and animals.

Breathing is the most imperious of needs. We live by the air before we live by bread; and this happens of itself, without painful struggles, without costly labour, almost without our knowledge. We do not set out, armed for war, to conquer the air by rapine, violence, cunning, barter and desperate labour; the supreme element of life enters our bodies of its own accord; it penetrates us and quickens us. Each of us has his generous share of it without giving the matter a thought.

To crown perfection, it is free. And this will last indefinitely until an ever ingenious Treasury invents distributing-taps and pneumatic receivers from which the air will be doled out to us at so much a piston-stroke. Let us hope that we shall be spared this particular item of scientific progress, for that, woe betide us, would be the end of all things: the tax would kill the tax-payer!

Chemistry, in its lighter moods, promises us, in the future, pills containing the concentrated essence of food. These cunning compounds, the product of our laboratories, would not end our longing to possess a stomach no more burdensome than our lungs and to feed even as we breathe.

The plant partly knows this secret: it draws its carbon quietly from the air, in which each leaf is impregnated with the wherewithal to grow tall and green. But the vegetable is inactive; hence its innocent life. Action calls for strongly flavoured spices, won by fighting. The animal acts; therefore it kills. The highest phase, perhaps, of a self-conscious intelligence, man, deserving nothing better, shares with the brute the tyranny of the belly as the irresistible motive of action.

But I have wandered too far afield. A living speck, swarming in the paunch of a grub, tells us of the brigandage of life. How well it understands its trade as an exterminator! In vain does the Crioceris-larva take refuge in an unassailable casket: its executioner makes herself so small that she is able to reach it.

Adopt such precautions as you please, you pitiable grubs, pose on your sprigs in the attitude of a threatening Sphinx, take refuge in the mysteries of a box, arm yourself with a cuirass of dung: you will none the less pay your tribute in the pitiless conflict; there will always be operators who, varying in cunning, in size, in implements, will inoculate you with their deadly germs.

Not even the lily-dweller, with her dirty ways, is safe. Her grub is as often the prey of another Tachina, larger than that of the Field Crioceris. The parasite, I am convinced, does not sow her eggs upon the victim so long as the latter is wrapped in its repulsive great-coat; but a moment's imprudence gives her a favourable opportunity.

When the time comes for the grub to bury itself in the ground, there to undergo the transformation, it lays aside its mantle, with the object perhaps of easing itself when it descends from the top of the plant, or else with the object of taking a bath in that kindly sunlight whereof it has hitherto tasted so little under its moist coverlet. This naked journey over the leaves, the last joy of its larval life, is fatal to the traveller. Up comes the Tachina, who, finding a clean skin, all sleek with fat, loses no time in dabbing her eggs upon it.

A census of the intact and of the injured larvæ provides us with particulars which agree with what we foresaw from the nature of their respective lives. The most exposed to parasites is the Field Crioceris, whose larva lives in the open air, without any sort of protection. Next comes the Twelve-spotted Crioceris, who is established in the asparagus-berry from her early infancy. The most favoured is the Lily-beetle, who, while a grub, makes an ulster of her excretions.

For the second time, we are here confronted by three insects which look as if they had all come out of one mould, so much are they alike in shape. If the costumes were not different and the sizes dissimilar, we should not know how to tell one from another. And this pronounced resemblance in figure is accompanied by a no less pronounced lack of resemblance in instinct.

The evacuator that soils its back cannot have inspired the hermit living in cleanly retirement inside its globe; the occupant of the asparagus-berry did not advise the third to live in the open and wander like an acrobat through the leafage. None of the three has initiated the customs of the other two. All this seems to me as clear as daylight. If they have issued from the same stock, how have they acquired such dissimilar talents?

Furthermore, have these talents developed by degrees? The Lily-beetle is prepared to tell us. Her grub, let us suppose, once conceived the notion, when tormented by the Tachina, of making the stercoral slit open above. By accident, with no definite purpose in view, it emptied the contents of its intestine over its back. The natty Fly hesitated in the presence of this filth. The grub, in its cunning, recognized, as time went on, the benefit to be derived from its poultice; and what at first was an unpremeditated pollution became a prudent custom.

As success followed upon success, with the aid of the centuries, of course, for these inventions always take centuries, the dung overcoat was extended from the hinder end to the fore-part, right down to the forehead. Finding itself the gainer by this invention, setting the parasite at defiance under its coverlet, the grub made a strict law of what was an accident; and the Crioceris faithfully handed down the repulsive great-coat to her offspring.

So far this is not so bad. But things now begin to become complicated. If the insect was really the inventor of its defensive methods, if it discovered for itself the advantage of hiding under its ordure, I look to its ingenuity to keep up the tricks until the precise moment has come for burying itself. But, on the contrary, it undresses itself some time beforehand; it wanders about naked, taking the air on the leaves, at a time when its fair round belly is more than ever likely to tempt the Fly. It completely forgets, on its last day, the prudence which it acquired by the long apprenticeship of the centuries.

This sudden change of purpose, this heedlessness in the face of danger tells me that the insect forgets nothing, because it has learnt nothing, because it has invented nothing. When the instincts were being distributed, it received as its share the overcoat, of whose methods it is ignorant, though it benefits by its advantages. It has not acquired it by successive stages, followed by a sudden halt at the most dangerous moment, the moment most calculated to inspire it with distrust; it is no more and no less gifted than it was in the beginning and is unable in any way to alter its tactics against the Tachina and its other enemies.

Nevertheless, we must not be in a hurry to attribute to the garment of filth the exclusive function of protecting the grub against the parasite. It is difficult to see in what respect the Lily-grub is more

The Glow-Worm and Other Beetles

deserving than the Asparagus-grub, which possesses no defensive arts. Perhaps it is less fruitful and, to make up for the poverty of the ovaries, boasts an ingenuity which safeguards the race. Nor is there anything to tell us that the soft coverlet is not at the same time a shelter which screens a too sensitive skin from the sun. And, if it were a mere fal-lal, a furbelow of larval coquetry, even that would not surprise me. The insect has tastes which we cannot judge by our own. Let us end with a doubt and proceed.

May is not over when the grub, now fully-grown, leaves the lily and buries itself at the foot of the plant, at no great depth. Working with its head and rump, it forces back the earth and makes itself a round recess, the size of a pea. To turn the cell into a hollow pill which will not be liable to collapse, all that remains for it to do is to drench the wall with a glue which soon sets and grips the sand.

To observe this work of consolidation, I unearth some unfinished cells and make an opening which enables me to watch the grub at work. The hermit is at the window in a moment. A stream of froth pours from his mouth like beaten-up white of egg. He slavers, spits profusely; he makes his product effervescence and lays it on the edge of the breach. With a few spurts of froth the opening is plugged.

I collect other grubs at the moment of their interment and install them in glass tubes with a few tiny bits of paper which will serve them as a prop. There is no sand, no building-material other than the creature's spittle and my very few shreds of paper. Under these conditions can the pill-shaped cell be constructed?

Yes, it can; and without much difficulty. Supporting itself partly on the glass, partly on the paper, the larva begins to slaver all around it, to froth copiously. After a spell of some hours, it has disappeared within a solid shell. This is white as snow and highly porous; it might almost be a globule of whipped albumen. Thus, to stick together the sand in its pill-shaped nest, the larva employs a frothy albuminous substance.

Let us now dissect the builder. Around the oesophagus, which is fairly long and soft, are no salivary glands, no silk-tubes. The frothy cement is therefore neither silk nor saliva. One organ forces itself upon our attention: it is the crop, which is very capacious, and dilated with irregular protuberances that put it out of shape. It is filled with a colourless, viscous fluid. This is certainly the raw material of the frothy spittle, the glue that binds the grains of sand together and consolidates them into a spherical whole.

When the preparations for the metamorphosis are at hand, the stomachic pouch, having no longer to do duty as a digestive laboratory, serves the insect as a factory, or a warehouse for different purposes. Here the Sitares store up their uric waste products; here the Capricorns collect the chalky paste which becomes the stone lid for the entrance to the cell; here caterpillars keep in reserve the gums and powders with which they strengthen the cocoon; hence the Hymenoptera draw the lacquer which they employ to upholster their silken edifice. And now we find the Lily-beetle using it as a store for frothy cement.[1] What an obliging organ is this digestive pouch!

[1] This subject is continued in the essay on the Foamy Cicadella. Cf. *The Life of the Grasshopper*: chap. xx. — *Translator's Note*.

The two Asparagus-beetles are likewise proficient dribblers, worthy rivals of their kinswoman of the lily in the matter of building. In all three cases the underground shell has the same shape and the same structure.

When, after a subterranean visit of two months' duration, the Lily-beetle returns to the surface in her adult form, a botanical problem remains to be solved before the history of the insect is completed. We are now at the height of summer. The lilies have had their day. A dry, leafless stick, surmounted by a few tattered capsules, is all that is left of the magnificent plant of the spring. Only the onion-like bulb remains a little way down. There, postponing the process of vegetation, it waits for the steady rains of the autumn, which will renew its strength and make it burgeon into a sheaf of leaves.

How does the Lily-beetle live during the summer, before the return of the green foliage dear to its race? Does it fast during the extreme heat? If abstinence is its rule of life in this season of vegetable dearth, why does it emerge from underground, why does it abandon its shell, where it could sleep so peacefully, without the necessity of eating? Can it be need of food that drives it from the substratum and sends it to the sunlight so soon as the wing-cases have assumed their vermilion hue? It is very likely. For the rest, let us look into the matter.

On the ruined stems of my white lilies I find a portion covered with a scrap of green skin. I set it before the prisoners in my jars, who emerged from their sandy bed a day or two ago. They attack it with an appetite which is extremely conclusive; the green morsel is stripped bare to the wood. Soon I have nothing left, in the way of their regulation diet, to offer my famished captives. I know that all the lilies, native or exotic, the Turk's cap lily, or Martagon, the lily of Chalcedon, the tiger lily and many others, are to their taste; I do not forget that the crown imperial fritillary and the Persian fritillary are equally welcome; but most of these delicate plants have refused the hospitality of my two acres of pebbles and those which it is more or less possible for me to grow are now as tattered as the common lily. There is not a patch of green left on them.

In botany the lily gives its name to the family of the Liliaceæ, of which it is the leading representative. Those who feed upon the lily ought also, in the absence of anything better, to accept the other plants of the same group. This is my opinion at first; it is not that of the Crioceris, who knows more than I do about the virtues of plants.

The family of the Liliaceæ is subdivided into three tribes: the lilies, the daffodils and the asparaguses. Not any of the daffodil tribe suit my famishing prisoners, who allow themselves to die of inanition on the leaves of the following genera, the only varieties with which the modest resources of my garden have allowed me to experiment: asphodel, funkia, or niobe, agapanthus, or African lily, tritelia, hemerocallis, or day lily, tritoma, garlic, ornithogalum, or star of Bethlehem, squill, hyacinth, muscari, or grape-hyacinth. I record, for whom it may concern, this profound contempt of the Crioceris for the daffodils. An insect's opinion is not to be despised: it tells us that we should obtain a more natural arrangement by separating the daffodils farther from the lilies.

In the first of the three tribes, the classic white lily, the plant preferred by the insect, takes the chief place; next come the other lilies and the fritillaries, a diet almost as much sought after; and lastly the tulips, which the season is too far advanced to allow me to submit for the approval of the Crioceris.

The third tribe had a great surprise in store for me. The red Crioceris fed, though with a very scornful tooth, on the foliage of the asparagus, the favourite dish of the Field Crioceris and the Twelve-spotted Crioceris. On the other hand, she feasted rapturously on the lily of the valley (*Convallaria maialis*) and on Solomon's seal (*Polygonatum vulgare*), both of which are so different from the lily to any eye untrained in the niceties of botanical analysis.

She did more: she browsed, with every appearance of a contented stomach, on a prickly creeper, *Smilax aspera*, which tangles itself in the hedges with its corkscrew tendrils and produces, in the autumn, graceful clusters of small red berries, which are used for Christmas decorations. The fully-developed leaves are too hard for her, too tough; she wants the tender tips of the nascent foliage. When I take this precaution, I can feed her on the intractable vine as readily as on the lily.

The fact that the smilax is accepted gives me confidence in the prickly butcher's-broom (*Ruscus aculeatus*), another shrub of sturdy constitution, admitted to the family rejoicings at Christmas because of its handsome green leaves and its red berries, which are like big coral beads. In order not to discourage the consumer with leaves that are too hard, I select some young seedlings, newly sprouted and still bearing the round berry, the nutritive gourd, hanging at their base. My precautions lead to nothing: the insect obstinately refuses the butcher's-broom, on which I thought that I might rely after the smilax had been accepted.

We have our botany; the Crioceris has hers, which is subtler in its appreciation of affinities. Her domain comprises two very natural groups, that of the lily and that of the smilax, which, with the advance of science, has become the family of the Smilaceæ. In these two groups she recognizes certain genera—the more numerous—as her own; she refuses the others, which ought perhaps to be revised before being finally classified.

An exclusive taste for the asparagus, one of the foremost representatives of the Smilaceæ, characterizes the two other Crioceres, those eager exploiters of the cultivated asparagus. I find them also pretty often on the needle-leaved asparagus (*A. acutifolius*), a forbidding-looking shrub with long, flexible stems bearing many branches, which the Provençal vine-grower uses, under the name of *roumiéu*, as a filter before the tap of the wine-vat, to prevent the refuse of the grapes from choking up the vent-hole. Apart from these two plants, the two Crioceres refuse absolutely everything, even when in July they come up from the earth with the famishing stomachs which the long fast of the metamorphosis has given them. On the same wild asparagus, disdainful of the rest, lives a fourth Crioceris (*C. paracenthesia*), the smallest of the group. I do not know enough of her habits to say anything more about her.

These botanical details tell us that the Crioceres, which hatch early, in the middle of summer, have no reason to fear famine. If the Lily-beetle can no longer find her favourite plant, she can browse upon Solomon's seal and smilax, not to mention the lily of the valley and, I dare say, a few others of the same family. The other three are more favoured. Their food-plant remains erect, green and well provided with leaves until the end of autumn. The wild asparagus even, undaunted by the extreme cold, maintains a sturdy existence all the year round. Belated resources, moreover, are superfluous. After a brief period of summer freedom, the various Crioceres seek their winter quarters and go to earth under the dead leaves.

CHAPTER XVIII
THE CLYTHRÆ

The Lily-beetle dresses herself: with her ordure she makes herself a cosy gown, an infamous garment, it is true, but an excellent protection against parasites and sunstroke. The weaver of fæcal cloth has hardly any imitators. The Hermit-crab dresses himself: he selects to fit him, from the discarded wardrobe of the Sea-snail, an empty shell, damaged by the waves; he slips his poor abdomen, which is incapable of hardening, inside it and leaves outside his great fists of unequal size, clad in stone boxing-gloves. This is yet another example rarely followed.

With a few exceptions, all the more remarkable because they are so rare, the animal, in fact, is not burdened by the need of clothing itself. Endowed, without having to manufacture a thing, with all that it wants, it knows nothing of the art of adding defensive extras to its natural covering.

The bird has no need to take thought of its plumage, the furry beast of its coat, the reptile of its scales, the Snail of his shell, the Ground-beetle of his jerkin. They display no ingenuity with the object of securing protection from the inclemencies of the atmosphere. Hair, down, scales, mother-of-pearl and other items of the animal's apparel: these are all produced of their own accord, on an automatic loom.

Man, for his part, is naked; and the severities of the climate oblige him to wear an artificial skin to protect his own. This poverty has given rise to one of our most attractive industries.

He invented clothing who, shivering with cold, first thought of flaying the Bear and covering his shoulders with the brute's hide. In a distant future this primitive cloak was gradually to be replaced by cloth, the product of our industry. But under a mild sky the traditional fig-leaf, the screen of modesty, was for a long while sufficient. Among peoples remote from civilization, it still suffices in our day, together with its ornamental complement, the fish-bone through the cartilage of the nose, the red feather in the hair, the string round the loins. We must not forget the smear of rancid butter, which serves to keep off the Mosquito and reminds us of the unguent employed by the grub that dreads the Tachina.

In the first rank of the animals protected against the bite of the atmosphere without the intervention of a handicraft are those which go clad in hair, dressed free of cost in fleeces, furs or pelts. Some of these natural coats are magnificent, surpassing our downiest velvets in softness.

Despite the progress of weaving, man is still jealous of them. To-day, as in the ages when he sheltered under a rock, he values furs greatly for the winter. At all seasons he holds them in high esteem as ornamental accessories; he glories in sewing on his attire a shred of some wretched flayed beast. The ermine of kings and judges, the white rabbit-tails with which the university graduate adorns his left shoulder on solemn occasions carry us back in thought to the age of the cave-dwellers.

Moreover, the fleecy animals still clothe us in a less primitive fashion. Our woollens are made of hairs interlaced. Ever since the beginning, without hoping to find anything better, man has clothed himself at the expense of the hairy orders of creation.

The bird, a more active producer of heat, whose maintenance is a more delicate matter, covers itself with feathers, which overlap evenly, and puts round its body a thick cushion of air on a bed of down. It has on its tail a pot of cosmetic, a bottle of hair-oil, a fatty gland from which the beak obtains an ointment wherewith it preens the feathers one by one and renders them impermeable to moisture. A great expender of energy by reason of the exigencies of flight, it is essentially, chilly creature that it is, better-adapted than any other to the retention of heat.

The Glow-Worm and Other Beetles

For the slow-moving reptile the scales suffice, preserving it from hurtful contacts, but playing hardly any part as a bulwark against changes of temperature.

In its liquid environment, which is far more constant than the air, the fish requires no more. Without effort on its part, without violent expenditure of motor force, the swimmer is borne up by the mere pressure of the water. A bath whose temperature varies but little enables it to live in ignorance of excessive cold or heat.

In the same way, the mollusc, for the most part a denizen of the seas, leads a blissful life in its shell, which is a defensive fortress rather than a garment. Lastly the crustacean confines itself to making a suit of armour out of its mineral skin.

In all these, from the hairy to the crustaceous, the real coat, the coat turned out by a special industry, does not exist. Hair, fur, feather, scale, shell, stony armour require no intervention of the wearer; they are natural products, not the artificial creations of the animal. To find clothiers able to place upon their backs that which their organization refuses them, we must descend from man to certain insects.

Ridiculous attire, of which we are so proud, made from the slaver of a caterpillar or the fleece of a silly sheep: among its inventors the first and foremost is the Crioceris-larva, with its jacket of dung! In the art of clothing itself, it preceded the Eskimo, who scrapes the bowels of the seal to make himself a suit of dittos; it forestalled our ancestor the troglodyte, who borrowed the fur-coat of his contemporary the Cave-bear. We had not got beyond the fig-leaf, when the Crioceris already excelled in the manufacture of homespun, both providing the raw material and piecing it together.

For reasons of economy and easy acquisition, its disgusting method, but with very elegant modifications, suits the clan of the Clythræ and Cryptocephali, those pretty and magnificently coloured Beetles. Their larva, a naked little grub, makes itself a long, narrow pot, in which it lives just like the Snail in his shell. As a coat and as a dwelling the timid creature makes use of a jar, better still, of a graceful vase, the product of its industry.

Once inside, it never comes out. If anything alarms it, with a sudden recoil it withdraws completely into its urn, the opening of which is closed with the disk formed by the flat top of the head. When quiet is restored, it ventures to put out its head and the three segments with legs to them, but is very careful to keep the rest, which is more delicate and fastened to the back, inside.

With tiny steps, weighted by the burden, it makes its way along, lifting its earthenware container behind it in a slanting position. It makes one think of Diogenes, dragging his house, a terra-cotta tub, about with him. The thing is rather unwieldy, because of the weight, and is liable to heel over, owing to the excessive height of the centre of gravity. It makes progress all the same, tilting like a busby rakishly cocked over one ear. One of our Land-snails, the Bulimus, whose shell is continued into a turret, moves almost in the same fashion, tumbling repeatedly as he goes.

The Clythra's is a shapely jar and does credit to the insect's art of pottery. It is firm to the touch, of earthy appearance and smooth as stucco inside, while the outside is relieved by delicate diagonal, symmetrical ribs, which are the traces of successive enlargements. The back part is slightly dilated and is rounded off at the end with two slight bumps. These two terminal projections, with the central furrow which divides them, and the ribs marking additions, which match on either side, are evidence of work done in two parts, in which the artist has followed the rules of symmetry, the first condition of the beautiful.

The front part is of rather smaller diameter and is cut off on a slant, which enables the jar to be lifted and supported on the larva's back as it moves. Lastly, the mouth is circular, with a blunt edge.

Any one finding one of these jars for the first time, among the stones at the foot of an oak, and wondering what its origin could be, would be greatly puzzled. Is it the stone of some unknown fruit, emptied of its kernel by the patient tooth of the Field-mouse? Is it the capsule of a plant, from which the lid has dropped, allowing the seeds to fall? It has all the accuracy, all the elegance of the masterpieces of the vegetable kingdom.

After learning the origin of the object, he would be no less doubtful as to the nature of the materials, or rather of their cement. Water will not soften, will not disintegrate the shell. This must be so, else the first shower of rain would reduce the grub's garment to pulp. Fire does not affect it greatly either. When exposed to the flame of a candle, the jar, without changing shape, loses its brown colour and assumes the tint of burnt ferruginous earth. The groundwork of the material therefore is of a mineral nature. It remains for us to discover what the cement can be that gives the earthy element its brown colour, holds it together and makes it solid.

The grub is ever on its guard. At the least flurry, it shrinks into its shell and does not budge for a long time. Let us be as patient as the grub. We shall surely, some day or other, manage to surprise it at work. And indeed I do. It suddenly backs into its jar, disappearing inside entirely. In a moment it reappears, carrying a brown pellet in its mandibles. It kneads the pellet and works it up with a little earth gathered on the threshold of its dwelling; it softens the mixture as required and then spreads it artistically in a thin strip on the edge of the sheath.

The legs take no part in the job. Only the mandibles and the palpi work, acting as tub, trowel, beater and roller in one.

Once more the grub backs into its shell: once more it returns, bringing a second clod, which is prepared and used in the same manner. Five or six times over, it repeats the process, until the whole circumference of the mouth has been increased by the addition of a rim.

The potter's compound, as we have seen, consists of two ingredients. One of these, the first earth that comes to hand, is collected on the threshold of the workshop; the other is fetched from inside the pot, for, each time that the grub returns, I see it carrying a brown pellet in its teeth. What does it keep in the back-shop? Though we can scarcely find out by direct observation, we can at least guess.

Observe that the jar is absolutely closed behind, without the smallest waste-pipe by which the physiological needs from which the grub is certainly not immune can be relieved. The grub is boxed in and never stirs out of doors. What becomes of its excretions? Well, they are evacuated at the bottom of the pot. By a gentle movement of the rump, the product is spread upon the walls, strengthening the coat and giving it a velvet lining.

It is better than a lining; it is a precious store of putty. When the grub wants to repair its shell or to enlarge it to fit its figure, which increases daily, it proceeds to clean out its cess-pool. It turns round and, with the tips of its mandibles, collects singly, from the back, the brown pellets which it has only to work up with a little earth to make a ceramic paste of the highest quality.

Observe also that the grub's pottery is shaped like the legs of our peg-top trousers and is wider inside than at the opening. This excessive girth has its obvious use. It enables the animal to bend and turn when the contents of the cess-pit are needed for a fresh course of masonry.

A garment should be neither too short nor too tight. It is not enough to add a piece which lengthens it as the body grows longer; we must also see that it has sufficient fulness not to hamper the wearer and to give him liberty of movement.

The Snail and all the molluscs with turbinate shells increase the diameter of their corkscrew staircase by degrees, so that the last whorl is always an exact measure of their actual condition. The lower whorls, those of childhood, when they become too narrow, are not abandoned, it is true; they become lumber-rooms in which the organs of least importance to active life find shelter, drawn out into a slender appendage. The essential portion of the animal is lodged in the upper story, which increases in capacity.

The big Broken Bulimus, that lover of crumbling walls and limestone rocks leaning in the sun, sacrifices the graces of symmetry to utility. When the lower spirals are no longer wide enough, he abandons them altogether and moves higher up, into the spacious staircase of recent formation. He closes the occupied part with a stout partition-wall at the back; then, dashing against the sharp stones, he chips off the superfluous portion, the hovel not fit to live in. The broken shell loses its accurate form in the process, but gains in lightness.

The Clythra does not employ the Bulimus' method. It also disdains that of our dressmakers, who split the overtight garment and let in a piece of suitable width between the edges of the opening. To break the jar when it becomes too small would be a wilful waste of material; to split it lengthwise and increase its capacity by inserting a strip would be an imprudent expedient, which would expose the occupant to danger during the slow work of repair. The hermit of the jar can do better than that. It knows how to enlarge its gown while leaving it, except for its fulness, as it was before.

Its paradoxical method is this: of the lining it makes cloth, bringing to the outside what was inside. Little by little, as the need makes itself felt, the grub scrapes and strips the interior of its cell. Reduced to a soft paste by means of a little putty furnished by the intestine, the scrapings are applied over the whole of the outer surface, down to the far end, which the grub, thanks to its perfect flexibility, is able to reach without taking too much trouble or leaving its house.

This turning of the coat is accomplished with a delicate precision which preserves the symmetrical arrangement of the ornamental ridges; lastly, it increases the capacity by a gradual transfer of the material from the inside to the outside. This method of renewing the old coat is so accurate that nothing is thrown aside, nothing treated as useless, not even the baby-wear, which remains encrusted in the keystone at the original top of the structure.

If fresh materials were not added, obviously the jar would gain in size at the cost of thickness. The shell would become too thin, by dint of being turned in order to make space, and would sooner or later lack the requisite solidity. The grub guards against that. It has in front of it as much earth as it can wish for; it keeps putty in a back-shop; and the factory which produces it never slacks work. There is nothing to prevent it from thickening the structure at will and adding as much material as it thinks proper to the inner scrapings from the shell.

Invariably clad in a garment that is an exact fit, neither too loose nor too tight, the grub, when the cold weather comes, closes the mouth of its earthenware jar with a lid of the same mixed compound, a paste of earth and stercoral cement. It then turns round and makes its preparations for the metamorphosis, with its head at the back of the pot and its stern near the entrance, which will not be opened again. It reaches the adult stage in April and May, when the ilex becomes covered with tender shoots, and emerges from its shell by breaking open the hinder end. Now come the days of revelry on the leafage, in the mild morning sun.

The Clythra's jar is a piece of work entailing no little delicacy of execution. I can quite well see how the grub lengthens and enlarges it; but I cannot imagine how it begins it. If it has nothing to serve as a mould and a base, how does it set to work to assemble the first layers of paste into a neatly-shaped cup?

Our potters have their lathe, the tray which keeps the work rotating and implements to determine its outline. Could the Clythra, an exceptional ceramic artist, work without a base and without a guide? It strikes me as an insurmountable difficulty. I know the insect to be capable of many remarkable industrial feats; but, before admitting that the jar can be based on nothing, we should have to see the new-born artist at work. Perhaps it has resources bequeathed to it by its mother; perhaps the egg presents peculiarities which will solve the riddle. Let us rear the insect, collect its eggs; then the pottery will tell us the secret of its beginnings.

I install three species of Clythræ under wire-gauze covers, each with a bed of sand and a bottle of water containing a few young ilex-shoots, which I renew as and when they fade. All three species are common on the holm-oak: they are the Long-legged Clythra (*C. longipes*, FAB.), the Four-spotted Clythra (*C. quadripunctata*, LIN.), and the Taxicorn Clythra (*C. taxicornis*, FAB.).

I set up a second menagerie with some Cryptocephali, who are closely related to the Clythræ. The inmates are the Ilex Cryptocephalus (*C. ilicis*, OLIV.), the Two-spotted Cryptocephalus (*C. bipunctatus*, LIN.) and the Golden Cryptocephalus (*C. hypochoeridis*, LIN.), who wears a resplendent costume. For the first two I provide sprigs of ilex; for the third, the heads of a centaury (*Centaurea aspera*), which is the favourite plant of this living gem.

There is nothing striking in the habits of my captives, who spend the morning very quietly, the first five browsing on their oak-leaves and the sixth on her centaury-blooms. When the sun grows hot, they fly from the bunch of leaves in the centre to the wire trellis and back from the trellis to the leaves, or wander about the top of the cage in a state of great excitement.

Every moment couples are formed. They pester each other, pair without preliminaries, part without regrets and begin elsewhere all over again. Life is sweet; and there are enough for all to choose from. Several are persistent. Mounted on the back of the patient female, who lowers her head and seems untouched by the passionate storm, they shake her violently. Thus do the amorous insects declare their flame and win the consent of the hesitating fair.

The attitude of the couple now tells us the use of a certain organic detail peculiar to the Clythra. In several species, though not in all, the males' fore-legs are of inordinate length. What is the object of these extravagant arms, these curious grappling-irons out of all proportion to the insect's size? The Grasshoppers and Locusts prolong their hind-legs into levers to assist them in leaping. There is nothing of the sort here: it is the fore-legs which are exaggerated; and their excessive length has nothing to do with locomotion. The insect, whether resting or walking, seems even to be embarrassed by these outrageous stilts, which it bends awkwardly and tucks away as best it can, not knowing exactly what to do with them.

But wait for the pairing; and the extravagant becomes reasonable. The couple take up their pose in the form of a T. The male, standing perpendicularly, or nearly, represents the cross-piece and the female the shaft of the letter, lying on its side. To steady his attitude, which is so contrary to the usual position in pairing, the male flings out his long grappling-hooks, two sheet-anchors which grip the female's shoulders, the fore-edge of her corselet, or even her head.

At this moment, the only moment that counts in the adult insect's life, it is a good thing indeed to possess long arms, long hands, like *Clythra longimana* and *C. longipes*, as the scientific nomenclature calls them. Although their names are silent on the subject, the Taxicorn Clythra and the Six-spotted Clythra (*C. sexmaculata*, FAB.) and many others also have recourse to the same means of equilibrium: their fore-legs are utterly exaggerated.

Is the difficulty of pairing in a transversal position the explanation of the long grappling-irons thrown out to a distance? We will not be too certain, for here is the Four-spotted Clythra, who would flatly contradict us. The male has fore-legs of modest dimensions, in conformity with the usual rules; he places himself crosswise like the others and nevertheless achieves his ends without hindrance. He finds it enough to modify slightly the gymnastics of his embrace. The same may be said of the different Cryptocephali, who all have stumpy limbs. Wherever we look, we find special resources, known to some and unknown to others.

CHAPTER XIX
THE CLYTHRÆ: THE EGG

Let us leave the long-armed and short-armed to pursue their amorous contests as they please and come to the egg, the main object of my insect-rearing. The Taxicorn Clythra is the first in the field; I see her at working during the last days of May. A most singular and disconcerting batch of eggs is hers! Is it really a group of eggs? I hesitate until I surprise the mother using her hind-legs to finish extracting the strange germ which issues slowly and perhaps laboriously from her oviduct.

It is indeed the Taxicorn Clythra's batch. Assembled in bundles of one to three dozen and each fastened by a slender transparent thread slightly longer than itself, the eggs form a sort of inverted umbel, which dangles sometimes from the trelliswork of the cover, sometimes from the leaves of the twigs that provide the grub with food. The bunch of grains quivers at the least breath.

We know the egg-cluster of the Hemerobius, the object of so many mistakes to the untrained observer. The little Lace-winged Fly with the gold eggs sets up on a leaf a group of long, tiny columns as fine as a spider's thread, each bearing an egg as a capital. The whole resembles pretty closely a tuft of some long-stemmed mildew. Remember also the Eumenes' hanging egg,[1] which swings at the end of a thread, thus protecting the grub when it takes its first mouthfuls of the heap of dangerous game. The Taxicorn Clythra provides us with a third example of eggs fitted with suspension-threads, but so far nothing has given me an inkling of the function or the use of this string. Though the mother's intentions escape me, I can at least describe her work in some detail.

[1] Cf. *The Mason-wasps:* chap. i.—*Translator's Note.*

The eggs are smooth, coffee-coloured and shaped like a thimble. If you hold them to the light, you see in the thickness of their skin five circular zones, darker than the rest and producing almost the same effect as the hoops of a barrel. The end attached to the suspension-thread is slightly conical; the other is lopped off abruptly and the section is hollowed into a circular mouth. A good lens shows us inside this, a little below the rim, a fine white membrane, as smooth as the skin of a drum.

In addition, from the edge of the orifice there rises a wide membranous tab, whitish and delicate, which might be taken for a raised lid. Nevertheless there is no raising of a lid after the eggs are laid. I have seen the egg leave the oviduct; it is then what it will be later, but lighter in colour. No matter: I cannot believe that so complicated a machine can make its way, with all sail set, through the maternal straits. I imagine that the lid-like appendage remains lowered, closing the mouth, until the moment when the egg sees the light. Then and not till then does it rise.

Guided by the rather less complex structure of the eggs of the other Clythræ and of the Cryptocephali, I think of trying to take the strange germ to pieces; and I succeed after a fashion. Under the coffee-coloured sheath, which forms a little five-hooped barrel, is a white membrane. This is what we see through the mouth and what I compared with the skin of a drum. I recognize it as the regulation tunic, the usual envelope of any insect's egg. The rest, the little brown barrel, broached at one end and bearing a raised lid, must therefore be an accessory integument, a sort of exceptional shell, of which I do not as yet know any other example.

The Long-legged Clythra and the Four-spotted Clythra know nothing of packing their eggs in long-stemmed bundles. In June, from the height of the branches in which they are grazing, both of them carelessly allow their eggs to drop to the ground, one by one, here and there, at random and at long intervals, without giving the least thought to their installation. They might be little grains of excrement, unworthy of interest and ejected at hazard. The egg-factory and the dung-factory scatter their products with the same indifference.

Nevertheless, let us bring the lens to bear upon the minute particle so contumeliously treated. It is a miracle of elegance. In both species of Clythræ the eggs have the form of truncated ellipsoids, measuring about a millimetre in length.[2] The Long-legged Clythra's are a very dark brown and remind one of a thimble, a comparison which is the more exact inasmuch as they are dented with quadrangular pits, arranged in spiral series which cross one another with exquisite precision.

[2] .039 inch. — *Translator's Note.*

Those of the Four-spotted Clythra are pale in colour. They are covered with convex scales, overlapping in diagonal rows, ending in a point at the lower extremity, which is free and more or less askew. This collection of scales has rather the appearance of a hop-cone. Surely a very curious egg, ill-adapted to gliding gently through the narrow passages of the ovaries. I feel sure that it does not bristle in this fashion when it descends the delicate natal sheath; it is near the end of the oviduct that it receives its coat of scales.

In the case of the three Cryptocephali reared in my cages, the eggs are laid later; their season is the end of June and July. As in the Clythræ, there is the same lack of maternal care, the same haphazard dropping of the seeds from the centaury-blossoms and the ilex-twigs. The general form of the egg is still that of a truncated ellipsoid. The ornaments vary. In the eggs of the Golden Cryptocephalus and the Ilex Cryptocephalus they consist of eight flattened, wavy ribs, winding corkscrew-wise; in those of the Two-spotted Cryptocephalus they take the form of spiral rows of pits.

What can this envelope be, so remarkable for its elegance, with its spiral mouldings, its thimble-pits and its hop-scales? A few little accidental facts put me on the right track. To begin with, I acquire the certainty that the egg does not descend from the ovaries as I find it on the ground. Its ornamentation, incompatible with a gentle gliding movement, had already told me as much; I now have a clear proof.

Mingled with the normal eggs of both the Golden Cryptocephalus and the Long-legged Clythra, I find others which differ in no respect from the usual run of insects' eggs. The eggs are perfectly smooth, with a soft, pale-yellow shell. As the cage contains no other insects than the Clythra under consideration or the Cryptocephalus, I cannot be mistaken as to the origin of my finds.

Moreover, if any doubts remained, they would be dispelled by the following evidence: in addition to the bare, yellow eggs there are some whose base is set in a tiny brown, pitted cup, obviously the work of either the Two-spotted Cryptocephalus or the Long-legged Clythra, according to the cage, but unfinished work, which half-clothed the egg, as it left the ovaries, and then, when the dress-material ran short, or something went wrong with the machinery, allowed it to cross the outer threshold in the likeness of an acorn fixed in its cup.

Nothing could be prettier than this yellow egg, standing in its artistic egg-cup. Nor could anything tell us more conclusively where the jewel is manufactured. It is in the cloaca, the chamber common to the oviduct and the intestine, that the bird wraps its egg in a calcareous shell, often decorating it with magnificent hues: olive-green for the Nightingale, sky-blue for the Wheatear, soft pink for the Icterine Warbler. It is in the cloaca also that the Clythra and the Cryptocephalus produce the elegant armour of their eggs.

It remains to decide upon the material employed. From its horny appearance there is reason to believe that the little barrel of the Taxicorn Clythra and the scales of the Four-spotted Clythra are the products of a special secretion; and, now that it is too late, I much regret that I neglected to look for the apparatus yielding this secretion in the neighbourhood of the cloaca. As for the thing so prettily wrought by the Long-legged Clythra and the Cryptocephali, let us admit without false shame that it is made of fæcal matter.

The proof is furnished by certain specimens, by no means rare in the Golden Cryptocephalus, in which the customary brown is replaced by an unmistakable green, the sign of a vegetable pulp. In course of time, these green eggs turn brown and become like the others, no doubt by reason of an oxidization which alters the natural qualities of the digestive product still further. The egg, entering the cloaca in a soft and utterly naked state, receives an artistic coat of the intestinal dross, even as the Hen's egg is covered by a shell formed of the chalky secretions.

> *Materiem superabat opus, nam Mulciber illic*
> *Æquora celerat,*

said Ovid, in his description of the Palace of the Sun. The poet had precious metals and gems wherewith to build his imaginary marvel. What has the Clythra wherewith to achieve its ideal jewel? It has the shameful material whose name is banished from decent speech. And which is the Mulciber, the Vulcan, the artist-engraver that engraves the covering of the egg so prettily? It is the terminal sewer. The cloaca rolls the material, flutes it, twists it into spirals, decks it with chains of little pits and makes it up into a scaly suit of armour, showing how nature laughs at our paltry standards of value and how well able she is to convert the sordid into the beautiful.

In the bird, the egg-shell is a temporary defensive cell which at hatching-time is broken and abandoned and is henceforth useless. Made of horny matter or stercoral paste, the shell of the Clythra and the Cryptocephalus is, on the contrary, a permanent refuge, which the insect will never leave so long as it remains a larva. Here the grub is born with a ready-made garment, of rare elegance and an exact fit, a garment which it only has to enlarge, little by little, in the original manner described above. The shell, shaped like a little barrel or thimble, is open in front. There is nothing therefore to break, nothing to cast aside at the moment of hatching, except perhaps the actual envelope of the egg. Directly this membrane is burst, the tiny creature is free, with a handsome carved jacket, a legacy from its mother.

Let us indulge in a crazy dream and imagine young birds which keep the egg-shell intact, save for an opening through which they pass their head, and which, all their lives long, remain clad in this shell, on condition that they themselves enlarge it as they grow. This absurd dream is realized by our grub: it is dressed in the shell of its egg, expanded by degrees as the grub itself grows bigger.

In July all my collection of eggs are hatched, each isolated in a large cup covered with a slip of glass which will moderate the evaporation. What an interesting family! My vermin are swarming amid the miscellaneous vegetable refuse with which I have furnished the premises. They all move along with tiny steps, dragging their shells, which they carry lifted on a slant; they come halfway out and suddenly pop in again; they tumble over if they merely attempt to scale a sprig of moss, pick themselves up again, forge ahead and cast about at random.

Hunger, we can no longer doubt, is the cause of this agitation. What shall I give my famished nurselings? They are vegetarians: there can be no doubt whatever about that; but this is not enough to settle the bill of fare. What would happen under the natural conditions? Rearing the insects in cages, I find the eggs scattered at random on the ground. The mother drops them carelessly, here and there, from the top of the bough where she is refreshing herself by soberly notching some tender leaf. The Taxicorn Clythra fits a long stalk to her eggs and fixes them in clusters on the foliage. While I cannot yet make up my mind, in the absence of direct observation, whether the new-born larva cuts the suspension-thread itself, or whether the thread is broken merely as a result of drying up, sooner or later these eggs are lying on the ground, like the others.

The same thing must happen outside my cages: the eggs of the Clythræ and the Cryptocephali are scattered over the ground beneath the tree or plant on which the adult feeds.

Now what do we find under the shelter of the oak? Turf, dead leaves, more or less pickled by decay, dry twigs cased in lichens, broken stones with cushions of moss and, lastly, mould, the final residue of vegetable matters wrought upon by time. Under the tufts of the centaury on which the Golden Cryptocephalus browses lies a black bed of the miscellaneous refuse of the plant.

The Glow-Worm and Other Beetles

I try a little of everything, but nothing answers my expectations very positively. I observe, nevertheless, that a few disdainful mouthfuls are taken, a little bit here, a little bit there, enough to tell me the nature of the first layers which the grub adds to its natal sheath. With the exception of the Taxicorn Clythra, whose egg, with its suspension-stalk, seems to denote rather special habits, I see my several charges begin to prolong their shell with a brown paste, similar in appearance to that with whose manufacture and employment we are already familiar.

Discouraged by a food which does not suit them and perhaps also tried by a season of exceptional drouth, my young potters soon relinquish their task; they die after adding a shallow rim to their pots.

Only the Long-legged Clythra thrives and repays me amply for my troublesome nursing. I provide it with chips of old bark taken from the first tree to hand, the oak, the olive, the fig-tree and many others. I soften them by steeping them for a short time in water. The cork-like crusts, however, are not what my boarders eat. The actual food, the butter on the bread, is on the surface. There is a little here of all that the first beginnings of vegetable life add to old tree-trunks, all that breaks up decrepit age to turn it into perpetual youth.

There are tufts of moss, hardly a twelfth of an inch in height, which were sleeping droughtily under the merciless sun of the dog-days, but which a bath in a glass of water awakens at once. They now display their ring of green leaflets, brightened up and restored to life for a few hours. There are leprous efflorescences, with their white or yellow dust; tiny lichens radiating in ash-grey straps and covered with glaucous, white-edged shields, great round eyes that seem to gaze from the depths of the limbo in which dead matter comes to life again. There are collemas, which, after a shower, become dark and bloated and shake like jellies; sphærias, whose pustules stand out like ebony teats, full of myriads of tiny sacs, each containing eight pretty seeds. A glance through the microscope at the contents of one of these teats, a speck only just visible to the eye, reveals an astounding world: an infinity of procreative wealth in an atom. Ah, what a beautiful thing life is, even on a chip of rotten bark no bigger than a finger-nail! What a garden! What a treasure-house!

This is the best pasture put to the test. My Clythræ graze upon it, gathering in dense herds at the most luxuriant spots. One would take this heap for pinches of some brown, modelled seed or other, the snapdragon's, for instance; but these particular seeds push and sway; if one of them moves the least bit, the shells all clash together. Others wander about, in search of a good place, staggering and tumbling under the weight of the overcoat; they wander at random through that great and spacious world, the bottom of my cup.

Not a fortnight has elapsed before a strip, built up on the rim, has doubled the length of the Long-legged Clythra's shell, in order to maintain the capacity of the earthenware jar in proportion to the size of the grub, which has been growing from day to day. The recent portion, the work of the larva, is very plainly distinguishable from the original shell, the product of the mother; it is smooth over its whole extent, whereas the rest is ornamented with tiny holes arranged in spiral rows.

Planed away inside as it becomes too tight, the jar grows wider and at the same time longer. The dust taken from it, once more kneaded into mortar, is reapplied outside, more or less everywhere, and forms a rubble under which the original beauties end by disappearing. The neatly-pitted masterpiece is swamped by a layer of brown plasterwork; not always entirely, however, even when the structure reaches its final dimensions. If we pass an attentive lens between the two humps at the lower end, we very often see, encrusted in the earthy mass, the remains of the shell of the egg. This is the potter's mark. The arrangement of the spiral ridges, the number and the shape of the pits enable us almost to read the name of the maker, Clythra or Cryptocephalus.

From the very first I could not imagine the worker in ceramic paste designing its own pottery by drafting the first outlines. My doubts were justified. The grubs of the Clythra and the Cryptocephalus possess a maternal legacy in the shape of a shell, a garment which they have only to enlarge. They are born the owners of a layette which becomes the groundwork of their trousseau. They increase it, without, however, imitating its artistic elegance. A more vigorous age discards the laces in which the mother delights to clothe the new-born child.

Lightning Source UK Ltd.
Milton Keynes UK
UKHW010636301220
376134UK00001B/79